Julia M. Wright

Astronomy

The Sun and His Family

Julia M. Wright

Astronomy
The Sun and His Family

ISBN/EAN: 9783337119775

Printed in Europe, USA, Canada, Australia, Japan

Cover: Foto ©berggeist007 / pixelio.de

More available books at **www.hansebooks.com**

ASTRONOMY

The Sun and His Family

By JULIA MAC NAIR WRIGHT

Author of "Nature Readers, Seaside and Wayside,"
"Botany," etc.

ILLUSTRATED

" As heaven's high twins, whereof in Tyrian blue
The one revolveth in his course immense,
Might love his brother of the damask hue
For like and difference:

" For different pathways evermore decreed
To intersect, but not to interfere:
For common goal, two aspects and one speed,
One centre, and one year."

PHILADELPHIA

THE PENN PUBLISHING COMPANY

1898

CONTENTS

ASTRONOMY

CHAPTER I

THE FAIRY STORY OF THE SKIES

"This majestical roof, fretted with golden fire."

THE sun has set and a deep, rosy color lies along the western horizon. In the clear blue above this pink band a large star, with a soft, steady white light, shines out, the first of the evening host. Half-way up the eastern sky the moon has climbed, silver bright. The moonlight falls in a broad, glittering, bronze track across the waters.

In this line of light the rowboats and small sail-boats rock at anchor, each little fishing vessel with one steadying sail set; the trading schooners lie waiting for a favorable wind; they look like silver ships with silver sails. Such is the illusion of the moonlight.

We dwellers in the Occident and the North are

told that the true glory of night is never revealed to us; it is in the Orient and over the desert that the splendor of the heavens is fully shown, and there the starry hosts won their first worshipers, almost compelling man to adoration of the created and the seen, because of such surpassing glory.

Nowhere in our Western world can the heavens be more beautiful, more beguiling to our thoughts than near the sea, when the great, blue, scintillating dome bends over and mirrors itself in the wrinkling or foam-flecked waters, reproducing the starry lights a thousand fold. Of such a scene, Flammarion, the Romance Astronomer, writes : " The sky is studded with brilliant stars, the air is calm and slumberous, the silence of profound peace covers the earth, and in the smooth mirror of the waters the heavenly bodies are reflected, opening a new world before our eyes. Thought floats between two immensities; the infinite sky, the water, peopled also with stars."

In such beauty, when the uproar of the day has died almost to silence, fancy revels in the long, bright, wonderful, seemingly endless story of the starry hosts. We look at the white moon and consider how she draws the tides. The sea is brimming like a cup full to the overflow. It seems as if the waters rounded up like a bowl turned upside-down; they seem to rise toward the moon, and long to reach her. We almost see the silver chains by which she draws the floods to herself. But those silver chains no eye can see. Why does that serene planet covet these oceans? It is that the earth is the mother of the moon. Long ago the moon sprung out from the earth, as a child springs away from its mother's arms, and since then earth's truant moon-child has never returned. Yet twice in each twenty-four hours a great longing for her mother seizes the moon, and she draws with all her might to bring that mother to herself. As for six hours she pulls steadily with those silver chains known as "attraction" in our mortal speech, the waters feel the drawing and heap up on the side of the earth toward the moon. If the moon stood still and continued to. pull the seas, they might overflow all the astonished lands.; but always at the end of six hours, the moon becomes discouraged and relaxes her pull; she has moved so far along in her journey around the world that she is pulling other waves.

In this fairy story of the skies, however, we are but telling half facts, and half facts are dangerous; really the moon is pulling or attracting the earth all the twenty-four hours, and the waters are as constantly rising to meet her. In her trip around the world she exerts her power at its maximum on different places, and those are the places which are just beneath her; as she passes on her way the place of greatest attraction changes.

The moon attracts or draws the land as well as the water, but we do not notice this so much, so we speak freely of her drawing the tides and very seldom about her drawing the land. There is high tide twice each day on each side of the earth; the low tides are the half-way stations.

It is well to think high thoughts; it leads us to living higher lives. Let us study for a while the real story of the skies, not the fairy story, but the story that is true. Yet the real story is very like the fairy story. The moon was once a part of the earth. As the earth went whirling round and round, that force which is called tangential, or centre-flying force, caused a portion of the whirling ball to fly away. That flying part was the moon. As tangential force flung it off, after a while another force set to work and kept it from flying on, out of sight forever. This second force is called attraction. Every large body or mass will exert this force more or less strongly upon other bodies. The attraction of the earth holds the moon from moving farther away, and the attraction of the moon creates tides upon the earth.

Let us say that the case of the moon and the earth is like the case of a cup and ball. The ball

goes a certain distance, and the string hinders it
from going farther; the string represents attractive
force. If the cup is whirled by the handle, that
whirling acts upon the ball, and one can keep it at
the end of the tense string, whirling about the cup
from which it was flung. There are many points
where the cup and ball fail as an example of the
case of the earth and the moon, but let these do for
the present. Did the moon then break off in a great
lump, leaving a deep hole where it tumbled out of
the earth? Was that deep hole the hollow where
now the ocean runs? We are getting back to fairy-
tale talk; there was no lump, no hole. When the
moon flew from the earth the earth was not a solid,
not even a liquid; it was a great globe of gas,
thinner and hotter than anything we can imagine.

Suppose that when that flying-off happened some
being from afar, some angel or sprite, had been
looking on, would that one have felt wild surprise at
the evolutions of that moon-piece? Not at all. That
onlooker would have already seen many such violent dividings of matter; this forming of the moon
was far along in the series. What, then, was the
first one? Who can tell? Of that the wisest knows
nothing. All that any one knows is something, and
that a very little, of what happened in one little cor-
ner of the universe, in what is called our solar system.

We look up into the sky; since we began to think of this story of the skies the sunset color has faded from the west, the dome is darker blue and all aflame with myriad lights. A broad band of light lies like a ribbon across the heavens. We call it the Milky Way. It is a part of the heavens where the stars seem crowded together, as if they even touched each other, and in so seeming they mingle into one path of effulgence. In very truth they are far, far apart. It is only distance that makes them seem near; all heavenly bodies are far apart. This must be so because of tangential and attractive force. If they were too near the pull of attraction would cause the smaller body to fall into the larger; the strong pull would bring the smaller body out of its orbit. It must be just near enough to be held; not so near as to be drawn from its proper place. There must be a balance of tangential and attractive force.

Each heavenly body has its path in space, its appointed track, called its orbit. Sustained in space, these innumerable worlds travel their eternal ways and never interfere; each moves responsive to some sublime, unchanging, perfect law. To be able to discover, and in some degree interpret and appreciate such law, is one of the great privileges of the human mind.

The broad band called, from its white appearance, the Milky Way is one of many nebulæ with which space is set. The white appearance is caused by the immense number and distance of the stellar systems which form these clouds of splendor. We speak of stellar systems, and we speak of our solar system: here is a distinction without a difference, for every sun is a star, and every system is a stellar system. This vast nebulæ which crosses the sky like a white ribbon, is made up of many of such stellar systems as ours, but most of them in all probability far larger than our own. Each particle of that apparently closely sown star-dust is a sun, and the systems that surround these suns are so distant that we cannot discern them even as star-dust. Let us try to fix our minds on the thought that this nebulæ of the Milky Way is only one of myriads of nebulæ, each lying like an isle of light in the seas of space, or anchored on the bosom of that vast dark ocean, as the ships are at anchor here in the harbor. This Milky Way we understand is far from the largest of these isles of light, yet for a ray to flash from one extreme of its greater axis to the other would require seventeen thousand years; yet light flashes seventy thousand leagues a second! Truly, now, we are thinking great thoughts, thoughts which take away our breath and dazzle our eyes, and leave us feeling very insignificant. Yet, it is something to be able to reach such conceptions as these.

For convenience in study astronomers, from the earliest ages, have grouped together certain stars which come within one field of observation. Star-gazers have seen that between certain stars they could trace some rude outline of some known thing, as, for instance, a chair, a dipper, a crown. These stars were set together then as the constellation of the Dipper, Crown, Orion, etc., etc. We must, however, remember that the stars in a constellation may have no real relationship, no nearness except that which we have made for them, in fancy, and on our astronomical maps.

While a constellation is a thing arbitrary, and arranged for human convenience in study, a system is made up of bodies which really belong together, and have interdependence and mutual relationships. Thus our earth is part of the solar system, the centre of which is the sun. All the planets in this system have sprung from the sun, are held in their places by his attraction, are reached by his beams, and have mutual dependence and relations. The laws which govern this system are harmonious and invariable; they have been discovered, and their nature is such that with some degree of certainty we can assume that they are not limited to the solar system, but

extend through all space, and govern the most distant nebulæ. And how distant and how numerous
these are we cannot conceive. Also we have equal
reason to believe that "star stuff" is the same in all
nebulæ and all systems.

When we think back from distant to near,
what shall we seek for and what shall we find in
this little system of ours—the solar system—about
which we know that little something,—aggregate
of the researches, discoveries, deep-thinkings of the
wise of many ages? What is the life-story of the
solar system? How are worlds born, and how do
they grow? What is the story of the sun and his
family?

When we begin with the sun where shall we go
to find him? We must needs go to the Milky Way
to find the sun and his family. Here we trace
thought-ways over which many minds have passed
before. No doubt the heavenly bodies, distant,
mysterious, splendid, were among the first objects
which drew the studious thoughts of men. The
early years of the race were not the age of the bar-
barians, but of the thinkers. The thinking may
have had its child-like simplicity at first, but it
arrived at great results. Doubtless the first careful
observers of the skies were the shepherds, who
watched the flocks by night; the mariners, whose

little craft hugged the shores; those desert traders, the men of the caravans, who, resting in the heat of the noon-day, traveled on in the quiet coolness of the night. The rising and setting of the sun was matter of common observation to all classes, and it did not take long to learn that there was not a new sun every day, but that the same orb disappeared in the west, and rose again in the east. The farmers, no doubt, early interested themselves in the varied positions of the sun in the sky during the different periods of the year, marking the changes of the seasons. The mid-summer sun is high overhead at noon, but the mid-winter sun hangs low in the sky even at noon. Thus it was seen that the sun had a yearly as well as a daily movement.

The men whose business gave them opportunity to watch the skies at night, while the day laborers slept, slowly discovered other motions of the heavenly bodies; those especial stars, or those groups of stars which seemed to have some interdependence, also slowly journeyed toward a setting; and as the year passed on the positions of these constellations in the sky varied.

Year by year, as the long-lived, patient watchers of the skies recorded their observations, they saw that the sun had in his yearly journey one invariable path, moving through the domain of certain constel-

lations, and these they named the zodiac, and divided
into twelve signs, or stations, for the year. Then, as
the heavens must be supposed to dominate the
earth, by these zodiacal signs they governed their
affairs and employments.

Sir Robert Ball suggests that observations of the
journey of the moon among the stars, the discovery of
its times and phases were first understood, and the
knowledge of the sun's yearly motion came later.
But this was not all; these thoughtful observers of
the early time in their close searching of the skies
during wakeful night hours, had discovered that
there were five bright bodies, unlike all the others in
the blue dome; while other stars held their places, or
moved only as the whole concave seemed slowly to
wheel about, these five bright lights were wanderers
among the stars. They named them planets or rovers.

Slowly, for many ages observations were made and
theories were formed. Once, even the wisest had
thought the earth a great ocean-surrounded plain,
somehow upheld, around which the heavens revolved; the earth great, the heavenly bodies, its
attendants, smaller. In the second century A. D.
Ptolemy, the astronomer, produced a work which for
fourteen hundred years ruled the thinking of men in
all that concerned astronomy. Ptolemy had found
arguments to prove that the earth is a sphere; he saw

it poised in space ; he held that it was there fixed, and that all the starry hosts revolved about it. One great thinker said to him, " Do you not see that if it should be that this earth-ball revolved, then the apparent moving of the whole heavens could be accounted for ?"

The astronomer assented, but rejected the possibility that the earth did move; he held that it was absolutely fixed in its place, neither rotated nor pursued an orbit. The sun, moon, and planets moved, he said, and they moved in great circles, the proof of which was, that only circular motion could be perfect, and as the heavenly bodies could be imperfect in nothing, therefore each must pursue a circular path through space about the earth. Such was the early astronomical system of those who had arrived at no real conception of the laws and relationships of the sun and his family.

CHAPTER II

> "Heaven's ebony vault
> Studded with stars unutterably bright."

DAY is the time to study botany, geology, and many other sciences; solemn, restful night invites us to look upon the star-set skies, and review what the wise have said and noted after sunset.

We say "up" in the sky, and "down" here on earth, to begin with. Is that right? Perhaps some one on a far-off star says, "up" of here, and "down" of there. There is no up or down in space, in matter of fact. That is a form of expression, continued for convenience from the times when people believed that the earth was a vast plain, surrounded with water, and that the sky was over it, as a bowl may be turned upside down over a plate.

What we call "sky" is merely the limit of our vision in space. We speak of the horizon line where land and sky seem to meet; that is merely as far as we can see. Your eyes are better than mine, and are able to see farther off than mine. We both have a limit, determined by the rotundity of the earth.

2   17

Some certain place is twenty-five miles from us ; we cannot see it, because on the rounding of the earth it falls below our vision.

" I have seen mountains fifty miles off," says some one. True : they were so high that they came into view. You did not see the ground-line of those mountains.

Why are the skies blue ? we ask as we look above us. They are not blue. About our earth is an elastic envelope, from one hundred and twenty to two hundred miles thick ; we call it the atmosphere. The gas portions moving we know as air ; air violently driven we call wind, moving at different rates of speed. Thus while we cannot see our atmosphere except for the blueness, we can feel it. The blueness of what we call sky is caused by the depth of atmosphere into which we gaze. All clouds are within our atmosphere ; they are caused by the condensation of water which has risen in invisible vapor from the surface of our globe. As we watch these lovely clouds we say to ourselves, " If I had eagle's wings and flew above our atmosphere, would I find it all golden light ?" Then reason replies, " If you had such wings you could go no higher than the eagle. When you rose above a certain distance the air would become so thin, or rare, that you could not breathe. You would suffocate, as a fish does when

taken out of water—unless you first froze to death.''
Yes, we are dealing with fact, not fairy lore. Why
should we freeze to death when rising nearer the sun?

The little gain in nearness to the sun would not
make up to us for the loss of heat reflected from the
earth. The glass roof of a greenhouse admits sun heat
but does not readily allow it to pass out. It is "a trap
to catch a sunbeam." The atmosphere is our glass
roof, the heat poured by the sun on the earth is held
here by the atmosphere. Close to the earth's surface
the atmosphere is most dense, and retains most heat.
As we rise higher the atmosphere is thinner and less
heat is retained. Of course, if we could get half-way
to the sun let us say, we would find heat enough to
scorch us. All mountain tops are cold, often snow-
covered the year round, even in the tropics.

Let us fancy, however, that we could rise far beyond our atmosphere. The deep blue would slowly
darken—it would not change to golden light, but to
intense blackness. Darkness reigns in stellar space.
Our atmosphere is full of atoms of dust which seize
and reflect the light poured from the sun, or from any
light-reflecting object. By means of this reflection
and refraction light is rendered visible. If we wandered into starry spaces we should be lost in awful
darkness, unless we found something to serve, as
does our air and the dust it contains.

How many stars should we find in space? No one can tell. Those seen by the naked eye, although we call them countless, are only about five thousand, clearly visible. The telescope brings more into view.

Each of those stars is a sun, perhaps much larger, hotter, brighter, than our sun. Each of these suns is the centre of a system, each turning on its axis, and each journeying along some fixed path, carrying with it its planets, held by its attraction. Those planets we cannot see.

Beyond the stars which we see, with their systems wheeling about them, may be millions more of sun-stars, in a multitude and distance of which it startles the mind to think. It almost overwhelms us even to fancy so many rushing along, worlds on worlds, suns on "suns in golden splendor burning."

The machinery of the universe is perfect. Each sun, each system, moves in a fixed ordered path; these paths may intersect, but never interfere. God makes no mistakes, and His machinery never breaks down. We may learn something about these distant sun-stars by the use of the telescope and spectroscope; what we learn shows us that the same laws govern distant space that govern our system.

Where is our system in space? we ask as we look abroad. We find it hard to believe that we are a part of the Milky Way. Although that looks so far off, above us, and entirely out of our range, it is really our part of space, and in it our sun and his family find their home.

How many queries these skies suggest. In regard to the more distant stars, we may learn that they are single, double, or even triple ; that they are variable ; that they are of different colors : purple, red, blue, white, yellow, green. They can be weighed, but not measured. They are weighed by wonderful instruments, and by means of difficult calculations which we need not discuss in these simple studies. When all is told, we find that we know almost nothing about the far off sun-stars and their satellites.

Into our system come wanderers from afar, vagrants of the skies, we call them comets. They may have traveled through those black and silent spaces, and visited suns and systems which the most powerful telescopes have never brought into view. They may have been drawn by the attraction of those other suns for a time, and then, breaking away, have come to us. They come and go, and tell us nothing.

Meanwhile our sun holds in good order his well-governed family about him, and does not allow them to stray off in the reckless company of the comets. For this well-disciplined family, the sun generously pours out light and heat. He gives not only the

light and heat which they need and use, but so much
more, that, as a well-known astronomer says, "The sips
that a flying swallow takes from a river is as far from
exhausting the water in the river, as are the planets
from using all the light that streams from the sun."

We use the words "stars," "planets;" can we use
them interchangeably? Do they mean the same kind
of body? Not at all; a star and a planet differ widely.
Our sun is a star; if all stars are made of the same
"stuff" or matter as our sun, then the material that
makes planets is the same as that of which stars
are formed, for planets are made out of the material
of their central suns. The planet is distinguished
from a star by its disc and its motion also; planets
reflect light, but do not pour forth light as the sun-
stars. Several of the planets were, when first seen,
thought to be stars, until their motion through space
betrayed them to be individuals of our own solar
system. The most distant of our sister planets is
immensely nearer to us than the nearest star, and
thus even the farthest off of our planets reveals to
the telescope a disc, which the nearest star is too
distant to show through our most powerful telescopes.
Could we bring those far-off stars into as close view
as our sun, we should, no doubt, find them possessed
of all the characteristics of the sun, light-givers of
unutterable, unendurable splendor, immense, slowly

rotating, their envelopes rent and convulsed as is his, by leaping, struggling fires.

All this that the telescope reveals to us was unknown to those ancients, who, before history began to be recorded, gathered knowledge of the hosts of heaven simply as their unaided eyes could carry on observations. One does not then wonder that these first astronomers knew so little, but that they knew so much. It was much to detect those "sky wanderers," the five early known planets, and to name them and understand something of their motions.

The true theory of the universe was unknown until about the time of Columbus. Almost simultaneously the heavens and the earth revealed their secrets. The heavens unfolded to wrapt student minds the marvels of the solar system. The earth showed a new world, cradled on the waters near the gates of the sunset.

Copernicus is the next great light in astronomy in order of time after Ptolemy. Ptolemy took a great forward step in knowledge when he assured himself that the earth was a globe, but he left it fixed in space, the centre of a system, and the greatest body of that system. Copernicus established the great facts that the earth rotated on its axis, and that the seeming daily motions of the heavens was really due to the revolution of the earth itself. After this he assigned to the earth its true position in the universe: it was one of the planets revolving about the sun, one of the smallest of these planets. The sun was the shining centre of our system, and about him wheeled six planets, the five known to the ancients, Mercury, Venus, Mars, Jupiter, Saturn, and the Earth.

All this Copernicus demonstrated before a telescope was invented; that marvelous lens that should bring the distant near and introduce men to the wonders of the heavens was not yet invented. Simple dials and instruments for measuring the length of the shadow cast at noonday by the sun had been used in the calculations of the ancients. Three years after Copernicus died, Tycho Brahe was born, and on him the mantle of Copernicus seemed to have fallen. The King of Denmark built for Tycho Brahe an observatory, and there for twenty years this last of the astronomers without telescopes, studied the motions of the stars.

Galileo, born in Pisa, in 1642, was an astronomer with a happier destiny. When he was thirty-five years old the first astronomical telescope was constructed. In that square, gray Torre Galileo on the low hill above Florence, Galileo, the father of modern astronomy, wandered into new worlds until then "beyond our ken."

The telescope is an instrument fashioned on the principle of the human eye.   When the pupil of the eye is dilated to its utmost, then the most light is admitted, and the image cast on the retina has the greatest brilliancy.   Such objects as distant stars may not be bright enough to excite vision.   The telescope on its great lens catches beams which are too large to enter the human eye-pupil, and then concentrates these great rays into a strong, vivid, smaller ray, which excites the sensation of vision on the retina.

The larger the object glass in a telescope the more light it receives, and the greater is its power of revealing the hidden and bringing the distant near.

In looking at a telescope we are full of admiration for the scientific ingenuity which fashioned it, and with wonder that any valuable astronomical work was done without instruments, which now seem so indispensable.   It is the telescope which has explored for us the sun, his home, the number and manners of his family.

CHAPTER III

"The sun, the centre and the sire of light,
The keystone of the world-built arch of heaven."

"How can we know anything about the sun," one demands, "when it is too bright to be looked at?" The moon, by hiding the sun, helps us to study it. "How can the small moon hide the great sun? How can we study that which is hidden?" Look toward a house and hold, say, an oak leaf before your eyes. The oak leaf hides the house. It is not so much the relative bigness as the nearness of the one object to the eyes which enables it to hide another. The moon is so much nearer to the earth than the sun is that when, in its journey around the earth, the moon comes between the earth and the sun, the sun is hidden. This is called an eclipse.

There are three kinds of eclipses. When the moon comes between us and the sun so that the moon shadow crosses only one portion of the sun, that is a partial eclipse. When the shadow covers all the face of the sun, it is a total eclipse. When

26

a ring of sunlight is seen all around the moon-shadow, that is called an annular eclipse.

During a total eclipse of the sun the light begins to grow dim and birds and beasts seem uneasy. Looking through the smoked glass one may see a dark shadow, like a small part of a circle, creeping upon the glowing face of the sun. This black shadow enlarges, moving along the sun's face until that face is entirely covered by a circular shadow as large as the sun. Soon this begins to move beyond the sun's disc, and the part that had first been shadowed shows a margin of light. This increases until the last edge of the shadow slides away and the sun is as bright as ever.

A total eclipse of the sun affords the best opportunity for sun study. The sun being shadowed by the moon, all about the shadow we have a circle of brilliant, waving, flame-like light; this is called the corona. Beyond the corona leap up great spires and shafts of light, known as faculæ or torches. These are not different from the corona, but are the highest parts of it.

An eclipse lasts but a few minutes, and is not visible from all parts of the earth at the same time. When an eclipse is to take place astronomers hasten to the cities or other points from which it can best be seen. Various instruments are prepared, and when the eclipse occurs the observers are busy, some taking photographs, some studying the phenomena through telescopes, others using the spectroscope. When the eclipse is over the observers describe, compare, and discuss the various observations which they have made.

Around the edge of the sun are always to be seen prominences of brilliant burning matter, in great tumult. The light of the sun at ordinary times renders these torches nearly invisible, but during the obscurity of an eclipse they come out clearly. They are called heaps, jets, plumes; some leap up two hundred thousand miles from the sun's surface. These strange objects are masses of burning gas.

We have seen a sudden jet of flame start from some lump of burning coal, and with a puffing, hissing sound reach beyond the bars of the grate. From the burning central mass of the sun, vast jets of gas start up in flaming splendor of various colors. The surface of the sun is always in this commotion of burning plumes and peaks.

We have seen a prism held in a ray of sunlight; it divides the ray into its various colors; red, orange, yellow, green, blue, violet rays are cast upon a surface. An instrument called a spectroscope divides the light of the blazing sun plumes into their various colors. Different substances give forth vari-

ous colors of flame. We have noticed red, blue,
yellow, green flames in the grate. If you throw in
salt or soda you will change the color of the flames.
There are various colored flames in the grate because of various substances in the fuel. Each substance burns with its own proper color, and this
color does not vary.

When the spectroscope divides the colors of the
burning gas-spikes about the sun, and reflects upon
the screen the various colors for the observer, then
we certainly know what material is there burning in
the spike, for its color-tale is told and is reliable.

The spectroscope can also tell the color-tale of the
far-off stars. We burn the various substances which
compose our globe, and from the flames of their
gases we receive certain colors. We find these same
colors, divided by the spectroscope, from the burning sun spires, and we say that without a doubt the
sun is composed of the same elements that compose
our earth—for example, gold, iron, carbon, oxygen,
and so on, in the form of solids or gases.

The spectroscope and other instruments have
been so arranged that the sun can be observed and
studied at any time, but a total eclipse affords the
finest opportunity for sun study.

Not only has the sun its wonderful flame wreath,
shooting up into cones and banners of fire; it has

strange spots upon its surface. These spots vary as do the prominences. What the spots are has not been fully determined. Many astronomers conclude that they are rents or holes in the sun surface, through which the burning inner gas leaps out into the flames of the corona. It was by observation of these spots that the great fact of the turning of the sun upon its axis was discovered.

Now what facts do we know about the sun?

I. The sun is found to be the centre of our system.

II. It is, to the system, the source of light and heat.

III. Its attraction holds the planets of the system in their orbits.

IV. The sun is a globe, more than a million times larger than our earth.

V. The distance of the sun from our earth is over ninety-two millions of miles, a number greater than we can realize or imagine.

VI. The sun-spots have shown us that the sun is rolling over upon its axis. Our earth turns over once in twenty-four hours; the sun makes its revolution in twenty-five days.

VII. We have learned that while we say in general that the earth moves and the sun stands still, the sun really moves through space, **journeying** along an immense fixed orbit, carrying with it our entire solar system, with no variation of the relative distances of the planets.

VIII. Another grand fact learned is, that the sun and all the planets of our system are made of the same substances.

IX. The light of the sun is white light; its apparent yellow or golden tint is due to our atmosphere. From the white light we learn that the sun is not mere gas, nor is it a solid; it has probably a fluid crust, within which is gas; the sun is doubtless surrounded by an atmosphere of gas through which blaze up the flames of the corona and faculæ; these last vary so much that a spire may double its volume three or four times within the space of an hour.

The immense energy of the sun, causing these ceaseless changes, is a vast source of heat. The whole substance of the sun is in the highest possible state of heat. No fires of which we have any knowledge offer comparison to sun-combustion. It is true, however, that some good astronomers think that the sun-centre is cooled.

In respect to the other portions of the starry heavens, the sun is merely one of myriads of fixed stars. No doubt millions of these stars are greater and more brilliant than even our wonderful sun.

The laws that govern our system, the processes of the birth and growth of worlds here, doubtless hold good in all the other systems. From the one we know something of the many. "The sun is big enough for us at all events," we say in dog-days; "what would August be with a bigger sun?" Yes, we can say to our sun if his size is disparaged—

"Be pale afar, since still to me you shine!"

The worship of the sun was one of the earliest forms of idolatry. The Assyrians found among plants the oak tree, as a symbol of their sun-god, and among beasts the bull. To make the bull a more fit emblem of the sun-god they added to its images eagle's wings and a human head, thus uniting strength, swiftness, and wisdom, they had an image of the god who had hung the sun for his symbol in the heavens. A Persian fire-worshiper, whose chief divinity is the sun, once said to an Englishman: "The reason you northern people do not worship the sun is, because you have never really seen him. If he rose on you with the glory that he shows to us Persians, you would worship him."

Milton, in many passages of his poems, sung the matchless splendor of the sun; and as we cannot fail to realize, the advance of science, and our increasing knowledge of the "great luminary," serve to emphasize and give significance to such passages as—

"Thou sun, of this great world both eye and soul!"

or,

"At whose sight all the stars
Hide their diminished heads."

Far as our earth lies from the splendid source of its
heat, we should not be warmed by its rays were it
not for our atmosphere which spreads all about us,
softening and diffusing, and returning the solar
beams. Yet, if out of its course our earth strayed
a little nearer to the sun, it would shrivel in its heat,
as did the servants of Nebuchadnezzar, who approached to the seven-times heated furnace to throw
in the three Hebrew heroes.

When we speak of size we are at a loss for com-
parisons. For every acre on the surface of the earth
there are ten thousand acres of sun-surface. "If the
earth were represented by a mustard seed, then on
the same scale the sun should be represented by a
cocoanut," says one writer. If we could collect
and set to work sun heat, we should have no need
for coal, wood, gas, or electricity for heating. One
single square foot of sun surface pours forth heat
sufficient to produce continuous steam for a twenty
thousand horse-power engine. The heat is not lacking; all that is lacking is ability to grasp and apply
it. If, instead of the system which has sprung from
his mass and is governed by his attraction, our sun

had set around him in a "fiery ring" or coronal two thousand million globes such as our earth-home, he could provide warmth and light for them all. How childish was then that old idea that that splendid star of day came into being merely to nourish and equip this little world.

As far as we know the sun is the great sample spendthrift in existence. He sets his world-children an example of unsurpassable wasteful extravagance. He seems to squander his riches merely for the sake of lavish display. Off into space go light and heat, practically untaxed by the orbs of our system, so small is the portion received by them, in comparison with the amount distributed. What becomes of all that magnificent expenditure of solar radiation science has never discovered. It is, however, among the possibilities that science will yet so enlarge its powers that it may go into space after that apparently idly lavished largess, and find it not thrown away, but performing marvels of useful work.

When the greatest of multi-millionaires runs a course of boundless lavishment people say: "How will even millions stand that strain?" So one asks of the sun, "How is he going to keep up his expenditures?" As for fresh supplies of heat-giving material, Professor Langley states that the entire earth flung into the sun, as a stoker tumbles fuel into the fires under a great boiler, would not maintain the sun's lavishment for a single minute. All this has been going on for ages that cannot be numbered. What, then, keeps the heat up? Helmholtz discovered an answer for this question. The sun keeps up his own heat—by shrinking! If the sun were solid, by this time he would be cold and inert. If the sun were liquid he would have cooled to a tepid state and have no more heat than he needed at home; but the sun is largely gaseous, and being gaseous he daily contracts, and that contraction squeezes heat out of the sun, as pressure squeezes water out of a wet sponge. As long as this gaseous state continues and shrinkage goes on, our solar system will be well supplied with heat; and so vast is the volume of the sun that for incalculable ages this outpour may proceed without sensible diminution of light or heat.

CHAPTER IV

"Worlds on worlds are rolling ever
From creation to decay,
Like the bubbles on a river,
Sparkling, bursting, borne away."

LET us go back to the beginning of things, to the time when no worlds swung in space, far back to the time when there was nothing at all—even when there was no time! That is quite impossible; the mind cannot go back so far. Then let us go back to when there was nothing in our system; we will be satisfied with the less if we are denied the much. As far back as we can think, there was very much in our system. What we want to know is how the first planet began its course—the story of the sun's eldest child. Let us look far out beyond the low-hung moon and the nearer stars until we see among the constellations a thin, bright cloud, like a little, crumpled, tissue veil. Search the sky carefully to find more such faint luminous cloudlets.

Yes; one over there, one high up in mid-sky. Those are called nebulæ or clouds. To our eyes,

36

or to a telescope of low power, the nebulæ seem to be mere patches of dim haze. By means of a powerful telescope some of them will be found to be star clusters, so far off that their light unites, and we cannot see space between them. There are other nebulæ which the most powerful glasses cannot resolve into stars; they remain burning clouds. The spectroscope is set to write their story, and that shows that they are composed of gas, burning at a high degree of heat and tenuity, to which nothing known here can afford a comparison.

Some of the nebulæ seem to be turning on an axis. No one has yet settled the origin of the heat or the beginning of the spin of sun-spheres. It is not worth while to give here the various explanations suggested. Let us take it as proved that these nebulæ, composed of burning gas, show us the state of our system before the first planet was formed.

The sun is now more than a million times larger than our earth, and has probably a fluid crust. Let us go back to the time when the sun was many million times larger than now, hotter in proportion, and had no fluid crust. It was simply a vast globe of burning gas, revolving rapidly on its axis.

Let us use an illustration. If one whirls a wet mop, when it reaches a great speed the water begins to fly off. If you dip a wooden paddle into molasses, boiled to the candy point, and gathering up the thick molasses on the stick, begin to whirl it, as you whirl very fast, little lumps of the molasses begin to fly off. A velocity so great might be reached by machinery that the little lumps that flew off would themselves continue whirling about for a time.

This illustrates what we wish to explain, that this vast sun-sphere of glowing gas, spinning at a tremendous rate, went at last so fast that it could not hold together, and some of it was thrown off.

This portion did not go in a lump, or globe-like mass. It was thrown off around the equator, or great circle about the centre of the globe, the circumference where the diameter is greatest. The portion set free was a vast, luminous, gaseous ring.

Did you ever say that you were running so fast that you could not stop? It was so with this ring; it had gained so much velocity from the parent globe that it kept spinning on, in the same direction as the ball from which it had been freed.

Children are usually like their parents in appearance and manners. This first child of the sun was like the sun in substance and motion.

A ring cannot hold together as well as a ball; it is too large in extent for its thickness at any one point.

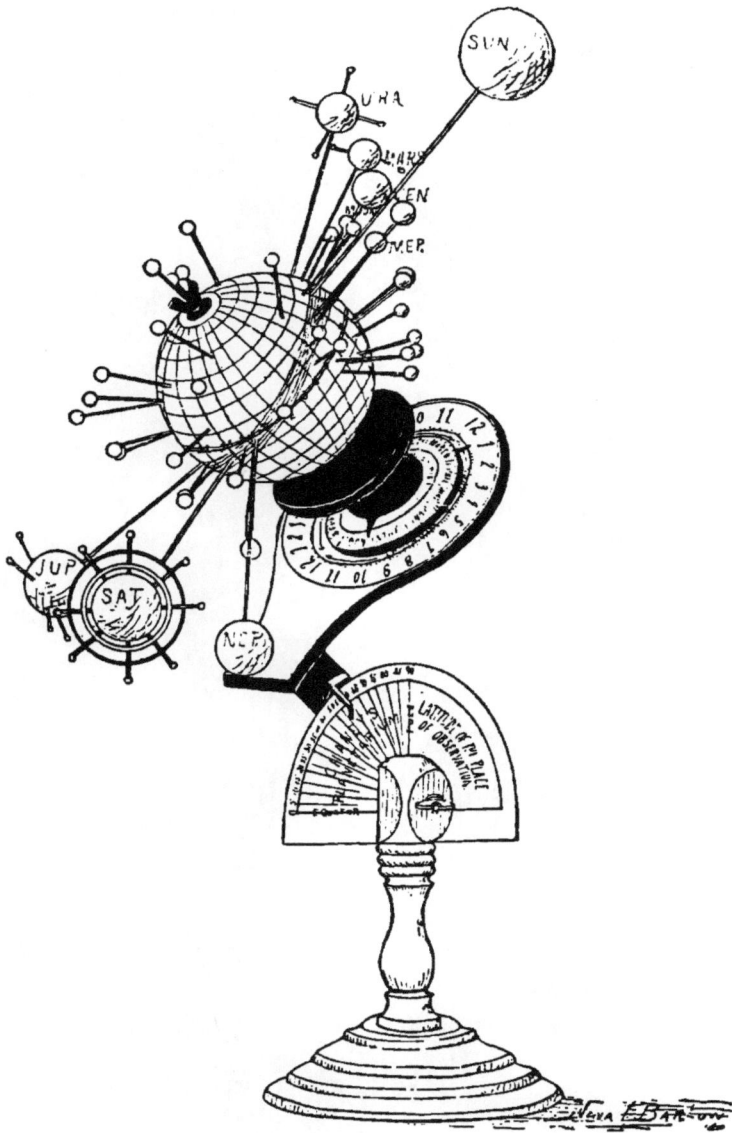

CHANEY'S PLANETARIUM

This ring thrown off from the sun broke up, and, by attraction of its parts, collected into a ball.

Stop a minute: let us settle matters as we go on. Why did not the parts tumble back into the sun by sun-attraction? Because, when first thrown off, the ring went so far from the sun that the attraction of the parts for each other was greater than the attraction of the sun over them, so they fell together. Thus in our system one planet was formed.

Why, when the ring broke, did not the several parts go on and on, each in its own direction? How could the parts collect themselves? We must consider first, that the ring, when it broke up, still had its parts remaining in the line and curve of the ring, and continuing the previous motion or journey. Lack of cohesion or attraction in the ring had permitted its division; the attraction of the original body prevented further departure from its centre. Any ring so dividing will have some parts nearer together than others; these will come together. A larger mass being formed by these unions, will exert more attraction and so collect more fragments.

The ring state is shown, as will be seen hereafter in Saturn; the coming together of broken rings into several satellites can be noted in Jupiter and other planets; the collection of all fragments into one satellite by the case of the earth and the moon.

The velocity of the sun-ball caused it to give the first planet such a tremendous fling when it was cast off that it went very far into space, was sent, in fact, to the outpost of our system. This new planet was like a grown-up lad, who becoming restless sets off on a journey to see the world, and to satisfy his eager curiosity goes as fast and as far as he can, while his father supplies his expenses and gives him his directions.

Our sun did not lose influence over his traveler in space. When the planet had gone to a certain distance, it could get no farther away; the attraction of the sun held it by invisible chains. Having reached its limit, it revolved on its axis, and traveled about the sun in a vast, far-off orbit, nearly circular, but slightly elongated, or, as it is called, elliptical. This is the shape of the orbit of all planets.

Why did not the sun-globe fling off a planet sooner? The reason was that as the sun-ball whirled, it cooled; cooling causes contraction or shrinkage, while contraction causes a swifter whirling. This process was needed before the point of throwing off a ring could be reached.

What was the name of this first planet, and where is it? It has been named Neptune, and is the out-

most planet in our system. It dwells on our bound-
ary line in space; it is our sentinel planet. Comets
which come flaming in, pass on their way this vast
fiery ball, which proclaims to them that they have
entered the dominions of the sun.

Why is Neptune fiery? One would think it might
have cooled by this time; it is countless ages since it
formed. It has grown cooler in the uncounted ages
since it found its place, but it is so vast and was cast
off with such velocity, at such an intense degree of
heat, that it is still a ball of burning gas. Neptune
is thirty times farther from the sun than the earth
is, so far off that from its surface the sun would appear as a great star. Owing to this great distance
from the source of light for this system, the stars
would be visible from Neptune day and night, if the
condition of its atmosphere rendered seeing anything
possible.

Neptune is one hundred times larger than our
earth. Its orbit is so vast that it takes one hundred
and sixty-five years to traverse it. Our earth completes her orbit in three hundred and sixty-five days.

No telescope yet made has been able to discover
features on the disc of Neptune; a very powerful instrument is needed to discover that it has a disc at all,
great fiery thing though it is! Neptune has thrown
off a ring of matter which has collapsed into a ball,

and this ball is Neptune's little moon, which travels around Neptune as our moon travels around our earth. The moon of Neptune moves at such marvelous speed that it makes its circuit in six days.

The planet Neptune was discovered in 1846. Before this date it had been occasionally seen through telescopes, but had been mistaken for a star until it had been searched for as a planet. A star does not change its station, and has no disc. In searching for a wanted planet, Neptune, then without a name and supposed to be a star, was discovered to have motion and a disc.

Why was a planet searched for?

Until then Uranus was supposed to be the outmost planet of our system. We know that while the sun by its attraction holds the planets in their orbits, planets influence each other by their attraction. It was found that Uranus was disturbed, at certain points in its orbit, as if swayed by the attractive force of some body other than the sun. Astronomers decided that there must be some great exterior planet and a search for it was rewarded in 1846 by the discovery of Neptune. This search for and finding of Neptune was one of the great triumphs of astronomy.

Why did they think the new planet was beyond Uranus?

A planet large enough to exert such attraction, would have been visible, if within the circle of Uranus. Neptune is very large, the third in size among the planets.

This discovery of the planet Neptune is a story of wonderful interest. It was no matter of accident; it was an affair of scientific research. Invariable laws ruling the planets had been discovered: for every perturbation or variation there must be a certain sufficient reason. A young Cambridge scholar, named Adams, having made careful studies of Uranus, saw that the presence of a planet exterior in the system would account for the various eccentricities of Uranus. He finally went to the Royal Observatory at Greenwich, and showed the Astronomer Royal, in charge there, the place in the skies where search should be made to find another child of the sun.

In the meantime Arago, a French astronomer, had called the attention of another famous French star-gazer, Le Verrier, to the singular behavior of Uranus, and Le Verrier also had made calculations, assuring himself that there must be an exterior planet, and indicating where it should be found. His calculations placed the desired planet only one degree from the place claimed for it by Adams. Here, then, was a very wonderful case; two astronomers pointed out the existence and almost the exact position of a planet that had never been seen!

Now was the time to begin a careful search. The method of search was this—an accurate chart was made of the part of the sky indicated as the residence of the distant unseen planet. Such charts of various parts of the heavens are now in the hands of many astronomers, but at that time no English astronomer had one. The making of such a chart involved much time and great labor, for the especial part of the sky had to be mapped out, all the visible stars in it carefully set down in their places, then all the stars brought to view by the telescope in that field must be added with the greatest accuracy, and then every night the sky thus mapped must be studied, and the stars in the sky compared with the stars on the chart to know if one of them was a rover. This chart work fell to Professor Challis. He marked out the region where the new planet was claimed to be, observed all the numerous bodies in that portion of sky and measured their distances. He used the great Northumberland Telescope, at Cambridge Observatory, England. He resolved after making his chart to go over it carefully, comparing star with star, and then a second and third time, and note if any vagrant had left his marked place. In truth Professor Challis did see Neptune and mark him as a star more than once, and a continuance of the observations would have assured him that this was the sun-child sought for.

In the meantime from France, Le Verrier had set the astronomers in Berlin and Paris at work planet hunting in a field of sky which he marked out. Several years previous one of the Berlin observers had made a star chart, and at the very time Le Verrier wrote to Berlin, asking for search for his planet, the chart of the particular part of the heavens which he indicated had just been engraved, but was not yet published. The day was September 23d; that night the sky was dazzlingly clear; the chart was laid out; Dr. Galle turned his telescope to the part of the blue vault indicated by Le Verrier's letter. One star after another was reviewed, identified. and found in its exact place on the chart. Finally Dr. Galle called out a clear, bright star of the eighth magnitude. "It is not on the chart," said the assistants. All was excitement, but astronomers need be careful, and perhaps there had been a mistake, and a star had been overlooked in making the chart. The men of the Berlin Observatory could scarcely wait for another night; it came brilliantly clear. Again the review began, star by star, and here again was this clear, glowing star of the eighth magnitude. But where? It had left its place; it had moved along an orbit; it had pursued the very path Le Verrier had foretold. They measured it; it had the exact diameter Le Verrier had asserted that it must have. This glowing jewel on the front of night had disc and motion, and shone with reflected, not self-outpoured, splendor. The child of the sun had been found. Here was the glorious reward of patient labor. For months Le Verrier and Adams, each unknown to the other, had fixed their minds on calculations and formulæ; they had reached human hands through space to grasp a planet on its tremendous journey; science had made its triumph; the planet was found.

CHAPTER V

"Look how the floor of heaven
Is thick inlaid with patines of bright gold."

"I WISH," we say when gazing toward a pale silvery patch among the far-off stars, "that I could see nebulæ plainer." We would doubtless find them very beautiful. Few objects surpass them when seen through the telescope. In addition, they lie at star beginnings. Let us know something more about them. Their state was once ours. They are named from the constellations within which they are seen. They may really be millions of miles away from these constellations, but, seen from the earth, they appear within their limits, and we so consider them for convenience. Some are called globe nebulæ and some ring nebulæ, from the shapes they take.

In the constellation of Taurus, or the Bull, is the nebula of the Crab; it is shaped like a great crab with extended claws. Sobieski's Crown is another strangely shaped nebula. In the constellation of Argo there is a nebula shaped exactly like a regula-

47

tion comet, with a burning head and a long, fiery tail. In the constellation of the Dog is a spiral nebula, a succession of spirals revolving upon each other in a blazing vortex.

Near the southern pole of the sky there are few stars visible, but there lie the Magellanic Clouds, two of the largest and most magnificent nebulæ, looking like a great storm of snow, where every flake is a patch of fire. The spiral nebula in the constellation of the Virgin is one of the most splendid objects upon which a telescope can be brought to bear. It is uplifting and enlarging to the mind to know of these glories in the sky, even if they lie beyond our sight.

The Milky Way itself is an enormous nebula in which are myriads of completed systems, and myriads more nebulæ which are yet whirling clouds of gas. The shape of the Milky Way is that of a wide ribbon, deeply notched at one end. What keeps these nebulæ from flying to atoms like dust and scattering about space?

Attraction. The force of attraction of part for part holds the parts from going beyond a certain limit, and by degrees draws them closer, until the vast revolving mass becomes a whirling ball. When Neptune was first formed no doubt it swiftly removed by tangential force to its present place, and then attractive force held it there. So of the other great planets. Their first place was their constant place. By such processes our system was formed.

What was the second planet of our system?

Uranus. Uranus was thrown off from the sun by the same process as Neptune, and next after Neptune. Uranus can only be seen through a strong telescope. It is sixty or seventy times larger than the earth, but is composed of so much lighter material that it is but fifteen times heavier. The distance of this planet from the sun is so great that it requires eighty-four years to complete its orbit.

Uranus has cast off a ring, or rings, of matter which broke into satellites, or moons. Four of these moons are known and named. The one nearest Uranus makes its circuit in two and one-half days, and the outmost one in thirteen days. A very strange fact about them is that they move from east to west, and not from west to east, as do all other known moons. Why this is no one has yet discovered. Sir Robert Ball says that in the early history of this planet there must have been local influences at work different from those which affected other bodies of our system.

At first more than four moons were assigned to Uranus, but some of these have been shown to be enormously distant stars, which chanced to enter

4

into telescopic view near the planet. We ask ourselves, "Why do the moons of Neptune and Uranus finish their circuit so much sooner than our moon does, when our planet is so much smaller?" It is because the orbit of our moon is really greater than theirs, as it is so far from our earth. The moons of Uranus and Neptune are very near their planets. Our moon was once much nearer our earth than now, but has slowly retreated. The nearness of the Uranian and Neptunian moons to these planets suggests that they may have been cast off at a comparatively recent period. Moons are at first near to their parent planet, and by tidal force drift away. On the other hand, the great planets remove at once to their orbits, and do not linger by the sun. Perhaps this difference is due to the greater vigor or force of the output.

When was Uranus found, and who found it? On the night of March 13th, 1781, an astronomer named William Herschel was studying the sky through a telescope that he had made for himself. On that night he found this planet. Herschel was an organist and music teacher, but so great a lover of astronomy that he devoted to that pursuit all his spare time and means. For eight years he had been carefully observing the sky, aided by his sister Caroline, who shared his passion for star-study. Herschel would

observe the stars through his telescope and detail
what he saw to Caroline, who sat by his side to write
out his observations. The succeeding day, while
Herschel was busy with his music teaching, his
sister worked out all the needed astronomical calculations.

Herschel had resolved to examine the sky with
care, and to map out all stars of a certain brightness.
He had directed his attention to the constellation of
Gemini, or the Twins, when he saw a star very different from the others. It had a disc—a very tiny
but true disc.

The eyes of Herschel were remarkably keen, and
his mental perceptions very quick. Other astronomers had seen this body and had called it a star; it
had changed its place and they had considered it
another star, not the same body in motion.

Night after night Herschel marked carefully this
wonder. At first he fancied it might be a comet.
But no; he soon saw that it was no comet, but a
body with a disc and moving in an orbit—a planet—
a new planet, lying far outside of Saturn, which had
until then been called the frontier orb of the solar
system.

This discovery made Herschel a famous astronomer. King George III became his friend, built an
observatory for him, and gave him a pension that

he and his sister might live near Windsor and devote themselves entirely to astronomy.

Herschel wished to name his new planet after King George, but the astronomers of the world objected to changing the ancient order of naming, and called the new planet Uranus, from the eldest of the gods of fable.

A careful study of the path of Uranus showed a vacillation and deviation in its motion. Uranus did not seem to be obeying the laws which govern our system—" Kepler's laws," as they are called, because he first clearly formulated and explained them. What was the trouble with Uranus? Was this second son of the solar family an unruly child? What was influencing him? He seemed to be paying attention not only to his parent sun, and to his younger brothers grouped between him and the sun, but to somebody beyond him. His attention was often seriously distracted. This became a matter of serious consideration.

In fact, Uranus was spending a part of his time corresponding with his elder brother, Neptune. At that time we, here upon earth, knew nothing about Neptune, but, of course, the sun understood all about him, and the devious ways of Uranus did not trouble him.

Astronomers set themselves to explain the erratic

and absent-minded conduct of Uranus, and, as was narrated in speaking of Neptune, this observation resulted in the discovery of Neptune in his inconceivably distant home. The planet of Uranus binds to immortal fame the name of the music-master-astronomer, William Herschel. He was one of the immortal few whose passion for star-study has ruled their lives, and the gratification of it has been more to them than meat and drink. If Herschel had not been famous as an astronomer, he might have been widely known as a remarkable maker of telescopes. The problem before him was this—he needed a telescope, but his means were inadequate to the purchase of even the mediocre instruments of his day. With indomitable energy he set about making a telescope. He was a man of extreme accuracy in all his undertakings, and as the result of several years labor he produced a very superior telescope. The fame of this instrument spread abroad, and various nobles and crowned heads requested him to make telescopes for them. By the sale of these he secured means for carrying on his own studies. The enthusiastic passion of Herschel for astronomy, his energy, persistency, his conquest of difficulties, all bring to mind an astronomer of the present day, Camille Flammarion, now world-famous. For him the stars are full of romance—so also is his scientific life full of romance.

A devotee of astronomy, not abundantly furnished with fortune, Flammarion, some fifteen years ago, had succeeded in getting together enough money to purchase a large telescope, but he had nowhere to mount it. The setting up of a telescope is about as costly as the instrument, for solidity in the foundation is required or the telescope is perverted by the jar. Professor Gavazzi Smythe once said that the great telescope in the observatory on the Calton Hill, Edinburgh, in spite of its isolation on the hill, and the immense piles of masonry upon which the instrument was placed, responded to the jar occasioned by wagons and drays rolling over the pavement down at the foot of the hill.

Flammarion concluded to ask his landlord to allow him to set up the telescope upon the roof of the house where he was then living. While he was considering plans and expense he received a very extraordinary letter signed E. Meret, Bordeaux. This letter contained an offer of a house, land, and money wherewith to set up a great private observatory at Juvisy, a village near Paris. Now the professor had often had practical jokes passed upon him by foolish folk who cannot understand why one should study the stars. He concluded this letter was another such joke, and did not reply to it. A second and a third of like tenor came. The third read in this

way : " I am more than seventy ; I am beginning to
lose my eyesight. But others live in the light and
know how to diffuse it. I repeat to you that I pos-
sess at Juvisy a small estate, where formerly I
dabbled with astronomy. That land I do not wish
to sell—I wish to present it to you. Its secular
shades will prove to you an oasis of blest repose.
Only answer me by one word, ' yes.' You will then
go and see it. If you do not like it, then you will
sell it."

It was just at this time that Flammarion's work,
" Popular Astronomy," leaped into unexpected and
unprecedented sale, and produced for him in a short
time $20,000. Here was money for the equipment
of his observatory, and an observatory was the desire
of his heart. Flammarion was now convinced that
the letter from Bordeaux was sent in good faith.
Instead of writing " yes " he went to Bordeaux and
saw Monsieur Meret. He found that the pretty little
estate in Juvisy, which was now made his own, was
called " Cour de France " or " French Court," because
in long-gone years the kings of France going from
Paris to Fontainebleau used to stop there while
their horses were changed. In this same house,
" Cour de France," Napoleon I slept the last night
before his abdication in 1814—probably not slept,
but remained awake, would be the fitter phrase.

Here at last arose on a low hill the tower of Camille Flammarion's observatory. Here he has gathered together beautiful astronomical and photographic instruments, and here he can make himself happy in his own especial way, studying the heavens and writing about them.

It is Flammarion's good fortune to be able to tell in graceful speech what he has seen and knows. Added to this fluent and elegant description are the illustrations of his subject, which photography enables him to make, and thus his work has done much to popularize astronomy.

Since the discovery of Uranus and the consequent discovery of Neptune, much of the modern astronomy is intensely mathematical and technical, and it is a boon to the popular mind when a descriptive writer such as Flammarion, dowered with enthusiasm and imagination, can do something to bring astronomy near to the general heart.

CHAPTER VI

"The planets in their stations listening stood."

" If we were flying through star space, what would
be the most splendid object that we should see?" This
is a question often asked by those watching the skies
on clear, brilliant nights.

We cannot possibly guess what is the crowning
glory of stellar space, but we can easily say what
would be the most splendid object within the solar
system, and it is hard to conceive that anywhere in
the heavens can be found a more glorious planet
than the third in order of the sun's children, next in
magnificence to the sun himself. We will assume the
sun's superior grandeur as a fact. The most wonderful sight within our solar system is the planet Saturn.

The five planets, Mercury, Venus, Mars, Jupiter,
Saturn, were known to the ancients, who named
them from their gods. They are visible to the naked
eye, and were called, as before said, " sky-wanderers,"
because they changed their places. They were
always understood to be different from the stars in
their nature.

57

All that we know of these planets is the slow accumulation of facts, gathered by ages of observation. After telescopes were invented, closer study enabled people to have a better acquaintance with them. For many hundred years Saturn was considered the least interesting of the planets. It was supposed to be the outmost orb in our system.

The distance of Saturn from the sun causes its motion through its orbit to be much slower than that of the nearer planets. This vast distance from the source of light also causes Saturn to be less brilliant than the other four planets which we have named.

A noted astronomer says: "To me it has always seemed that Saturn is one of the three most interesting celestial objects visible to observers in northern latitudes. The other two are the great nebula in Orion and the star cluster in Hercules."

Observers in southern latitudes might add to these, or put in place of one of them, the constellation of the Southern Cross.

The superior magnificence of Saturn cannot be known to the unaided eye. Seen without the telescope it is merely an orb more or less bright, according to its distance from us as it traverses its orbit.

At the farthest point of his orbit Saturn is some thousand millions of miles from the sun. The time required for his journey through this great path is over twenty-nine years.

Although Saturn is seven hundred times larger than our earth, it whirls so swiftly upon its axis that it turns over once in ten and one-half hours, while our earth requires twenty-four hours for one revolution. On Saturn the days would be but half as long as ours, but the seasons would be each about seven years long, instead of four months!

Saturn is the centre of a wonderful and complicated system of his own. This system occupies a space of four and one-half millions of miles diameter.

Saturn's system has been built out of Saturn, as the solar system has been built out of the sun.

We remember that when we speak of world building we begin with a whirling, burning globe, which casts off a burning ring. The great glory of Saturn is that he has continued to cast off rings, and these have not all collapsed into moons. Some of them are yet visible, through the telescope, as rings.

We must talk a little more about world building, in order the better to understand the wonders of Saturn. Each planet having been cast from the sun surface, receives from him its motion, and revolves upon its axis in the same direction as the sun turns upon his axis, because the sun gave the planet that impulse before it parted company with him.

The first form of the cast-off portion being the ring, the orbit of a planet continues to be nearly circular. The plane, or level, on which it moves, closely coincides with that of the equator of the sun, because the matter was originally cast off from that region. The simple experiment of whirling dough or thick tar upon the end of a stick will show us that when a ring of stuff flies off, it goes from the central portion of the mass, and so when stuff flies off from the sun to make any planet, or from a planet to make a moon, the portion which in whirling has bulged is the equatorial zone, so matter goes off from the central, or tropical part, midway between the poles. Let us fancy a vast nebula of irregular form; the parts attracting each other. By the law of gravitation these parts settle toward a centre, or may be to several centres, if parts with equally great attractive force are found. If, in a nebula there are several such points of attraction, several suns would be formed, and consequently their several systems would be evolved.

The nebula resolves itself into suns by the forces of attraction and gravitation. The particles come together toward a centre. Some consider that because the particles rush upon the centre from different sides that gives the mass the initial or first twirl, just as to set a top spinning we take the stem

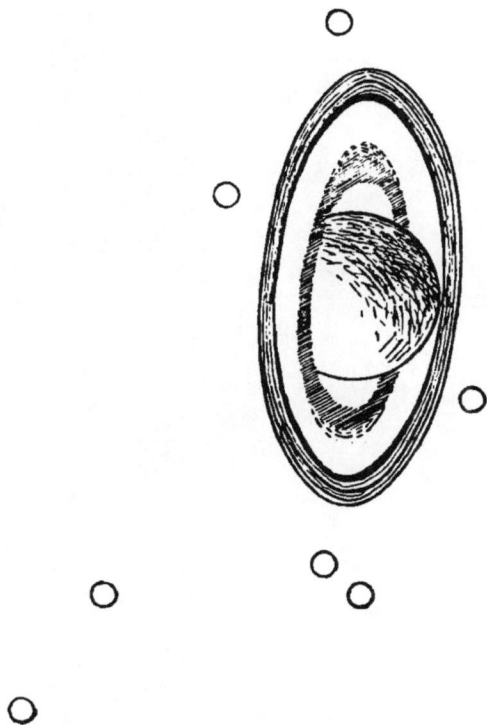

between our palms and pull one palm swiftly toward us, and push the other swiftly from us; this double motion creating a whirl which causes the top to spin.

The ball being formed and set spinning, its speed increases as shrinkage continues, and owing to this speed there is a heaping up, or tidal wave of material at the equator. When this heaping up reaches a point where it can overbalance gravitation, or the pull toward the centre, a ring of surface-matter will detach itself and remain poised. This ring will revolve in the same plane and direction as the body from which it parted, because it received both motion and material from that body.

When Saturn reached the point of casting off rings, the first ring or rings flung into space finally broke into moons.. Saturn is the richest in moons of any planet, having eight; some of these are very small.

Saturn then cast off other rings which have not broken up. If we stood upon Saturn and looked up, we would see two magnificent arches, bright·as the crescent moon, bending from horizon to horizon— a pair of great gleaming bows, revolving with terrible swiftness. Were it not for this swift rotation, causing the parts to cohere, the rings would go to pieces by force of gravitation.

"Wouldn't it be magnificent, if we had such rings!" we exclaim, rashly. It might have its disadvantages; for fifteen years at a time these rings must turn their dark sides to the planet which they span, and the regions below them would be hid in fearful night. Of what are Saturn's rings made? is a frequent query, the question meaning in what state is the ring material, for it is clear that it is of the same matter as the planet. Saturn is not so tenuous as Uranus or Neptune. It seems to have contracted to a state a little less dense than water. If the ball of Saturn could be dropped into an ocean, the ball would float with one quarter above water. Astronomers have suggested all manner of material for Saturn's rings, from big rocks to hydrogen gas. The spectroscope having investigated sun material, informs us of the material of which by consequence Saturn is made, but we do not know the degree of density of the rings. There is abundant room for more knowledge!

The surface of Saturn is probably a foggy envelope around a very hot interior. The rings are no doubt composed of the same gases, further solidified by cooling and contracting. The rings rotate in ten and a half hours. We have spoken of two rings; there are really three, perhaps four. The two outer rings are divided by a dark line, which probably represents intervening space. The inner ring is so hazy that it is called "the crape ring."

Now we can see Saturn with the vision of the mind. It is a great bright globe, whirling over and over, in a bright triple ring, as a ball in a hoop. Outside the great rings are eight moons, spinning over and traveling about Saturn, while Saturn travels millions of miles around the sun, carrying all this gorgeous company with him! This third son of the family has set up his own household. Let us think of Saturn's rings as great shining shoals of meteors or tiny moons, held together, whirling over, sweeping with Saturn through space.

The great interest which attaches to the planet Saturn is comparatively modern, because his marvelous system was not discovered until the telescope was brought to bear upon it. Neither was all this complicated object at once understood. The astronomer Galileo was the enchanter, and his little reflecting telescope of thirty diameters magnifying power was the potent wand which revealed the real Saturn to a wondering world. One can scarcely imagine the immense joy of Galileo, when turning his glass upon the sun he saw hitherto undreamed of spots on that splendid orb; when the moon revealed to him mountains on her silvery surface; when Jupiter displayed his satellites, and Venus showed a crescent like a new

moon! Then, full of joyful excitement, the lonely
watcher on the little gray " torre Galileo " turned his
telescope toward Saturn. Other astronomers with
telescopes had already seen, as he had, the wonders
in Sun, Moon, Venus, Jupiter; a discovery remained
for the good Galileo. What was this? Saturn appeared not one body, but three—three bodies in line,
always touching, always retaining the same relative
places; a great central body, and a smaller body on
the west, and one on the east. Sight so strange had
never before met human eyes, and Galileo dared not
speak of what he saw, until by further study he
could better comprehend the marvel. Of all men
the astronomer must be the most patient and per-
sistent. Week after week for two years Galileo
watched this triple Saturn—and saw the two outer
and smaller bodies slowly dwindle away! He had
during these two years asserted that Saturn was a
triple body, now it swam in space a single body, like
Jupiter, and he was compelled to admit it; his
enemies and envious friends gave a howl of derision.
When one looks at a photograph taken from a por-
trait painted of Galileo during his latter years, one
seems to see in the many lines the care, sorrow, and
mortification heaped up by this contumely and doubt;
while in the eyes the light of his final triumph burns
steadfastly. He wrote : " I do not know what to say

in a case so surprising, so unlooked-for, so novel.
Are those lesser stars consumed after the manner of
solar spots? The unexpected nature of the event
has greatly confounded me. Am I mistaken?"

Still the anxious astronomer watched the planet—
and back came the lesser orbs to view! More powerful telescopes were then constructed, and not Galileo, but all the objectors, began a careful study.
Presently they saw not two globes, one on each side
of Saturn, but two shining crescents, as if the golden
globe of Saturn had two vast crescent handles.
After seven or eight years these crescents disappeared, to return again and go through the same
changes as before.

Fifty years of observation were needed before the
real nature of these appearances was settled. In
1655 Huyghens, using a powerful lens, discovered
the shadow of a great ring cast upon the surface of
Saturn. What had been seen then were portions of
a mighty ring, which portions assumed the shape
of orbs, or crescents, or arms, according to the angle
at which they were observed.

Huyghens announced that Saturn was rotating on
its axis and surrounded by a vast ring, also rotating,
and so thin that when the edge was turned earthward it became invisible. It was nearly four years
before Huyghens completed the observations and

5

calculations which explained the Saturnian system. He then announced, "Saturn is girt with a ring, tenuous, flat, distinct from its surface, rotating, and inclined to the ecliptic." Thus we see that at first the ring of Saturn was supposed to be single.

Ten years passed before Cessini, another Italian astronomer, discovered that the ring of Saturn was not single, but double, was, in fact, two rings separated by a space, represented to the observer by a dark line. Later it was seen that the system of Saturn included three rings. In 1850 Professor Bond, of Cambridge, discovered a third ring, called "the crape ring," from its filmy texture. This discovery has led to the suspicion that the rings of Saturn, in their rotation, may be suffering change, and may be dividing themselves. A positive proof of the divisions between the rings would be the observation of a star in or through the intervening space. Only a large star would thus be visible, and so far none has been detected in passing. The outline of Saturn itself has been discovered through the intervening space between the first and second rings.

Majestic, in the distant skies, Saturn sweeps along its immense orbit, carrying the triple ring and four attendant moons, not the largest but the most splendid and complicated among the dependencies of the sun.

CHAPTER VII

THE INCREASE OF THE SUN FAMILY

"Night comes, world jewelled . . .
The stars rush forth in myriads, as to wage
War with the lines of darkness."

It is idle to discuss time in system building. No mind could estimate the endless procession of æons during which world after world, system after system, wheeled into order. Was all space once flaming, tumultuous, world building material? Did countless centres of gravity assert themselves, drawing material together until the infinite numbers of nebulæ were formed, and then, these dividing again by attraction of gravitation, and rolling and tumbling on, tornadoes of fire, system building came to be, so that in our nebula, the Milky Way, unnumbered systems shone, and other tumultuous centres were yet clouds of intense fire? Even in this one nebula, this one system of that world building, time is left out of account.

When, after the building of Saturn, the sun again had thrown off a ring of material, and that had come together into a sphere, it was by much the largest of the planets. This sphere, known to us as Jupiter, is the greatest of the sun family. It is between 1,200 and 1,500 times as large as our earth.* Its orbit is some 479,000,000 of miles from the sun, and therefore it receives much less light than our earth or Venus. Nevertheless, owing to its enormous size, it often shines more brightly than Venus, and seems truly the chief of the hosts of night. From this splendor of appearance the ancients gave it the name of the king of their gods.

If to Saturn you added all the other planets, you would not have a globe so large as Jupiter. The shape of Jupiter is more flattened than that of our earth. If you take an orange between your palms, the stem end in one palm, the blossom end in the other, and press your hands gently toward each other, you will enlarge the circumference about the centre or equator of the orange and produce a shape like that of Jupiter.

The orbit of Jupiter is more elliptical, or egg-shaped, than that of the earth. This orbit is 482,-000,000 of miles in extent, and the planet traverses it in twelve years, less fifty days. One rotation of Jupiter on its axis occupies nearly ten hours. Thus

---

*Ball says 1,200 times larger; Flammarion 1,500 times larger; Buritt ("Atlas of Astronomy") 1,300 times larger; Planets can be more accurately weighed than measured.

the day of Jupiter is less than half as long as ours, and each of his seasons is three years in duration. As far as we can judge, while we speak of seasons upon Jupiter, there would be really almost no change of temperature, neither summer nor winter, only perpetual spring.

Then dwellers on Jupiter would have neither snow, ice, nor torrid summer heat; endless spring might be wearisome! They would, however, encounter greater disadvantages than that. Jupiter, from its vast size, has retained a large part of its original heat, the heat of its particles when they divided from the sun. The surface of Jupiter has not cooled to solidity. How nearly solid it may be we cannot tell. Walking on Jupiter might be like walking upon the hot ashes lying on the sides of Vesuvius, shortly after an eruption—ashes which slip away and allow the feet to sink into the soft lava. Or, all the surface may be still boiling and bubbling as lava just poured from the crater. It may be even thinner than that, mere boiling mud and slime.

Another disadvantage inhabitants would encounter upon Jupiter would be the terrible storms. Jupiter is the most stormy planet known. The fiercest hurricanes, whirlwinds, cyclones on earth are gentle as compared to the tempests which ravage Jupiter. The surface of Jupiter and its atmosphere are constantly convulsed with tornadoes. On the whole, we conclude that there are no inhabitants upon Jupiter, although there are those who hold that life is possible upon all the planets. If Jupiter is so cloudy and stormy all the time, how can anything be learned about him, we inquire. The answer to this question may be unexpected; we learn much of his life-story by the observation of his satellites. Jupiter has cast off rings of matter which have contracted into five moons. These moons are of nearly equal size, and through a telescope they look like five small stars. We discover their true nature by finding that they are not fixed, as are the stars, but are ever roaming about a common centre, and that centre is Jupiter. They travel with him along his enormous journey through the skies.

The nearest of these satellites to the planet encompasses him in two days; the most distant in seventeen days. We see, in addition to a perpetual spring, Jupiter has the advantage of constant moonlit nights.

Now let us consider how these little moons have answered an important question in regard to Jupiter. Being so large and so bright a body, the question arose, " Does Jupiter receive all its light from the sun, as do the other planets, or is it itself a light giver?" When we see a moon of Jupiter going be-

tween the planet and the sun, we find that the great
planet is shadowed, because the moon cuts off the
sunlight from a portion of the surface of Jupiter.
This proves that the great orb is not a light-giver,
but a light-reflector, and receives its light from the
sun.

The moons of Jupiter give us another proof of
this fact. When Jupiter rolls between one of his
moons and the sun, the moon is lost to sight in darkness, because Jupiter cuts off from it the sunlight,
and has no light of his own to cast upon it.

One of the questions which arises when we observe
Jupiter is, "Why does Jupiter have such fierce
storms?" Storms on this earth are occasioned by
sun-heat. The sun's heat evaporates and raises
moisture from the earth's surface, which moisture
condenses into clouds and returns to the earth in
heavy rains or light showers and dews. Sun-heat
also causes a vacuum here and there by making the
air so light that it rises higher and higher. Then,
to fill the vacuum, colder currents of air rush in,
with greater or less velocity. These are called wind-
storms, hurricanes, tornadoes, and so on.

Jupiter receives too little sun-heat to account for
its storms. They are to be attributed to its own
heated condition. Just as the sun is always in a
fiery storm and tumult arising from his own terrible

heat, Jupiter is still hot enough in itself to create its own tempests.

While Jupiter is twelve or fourteen hundred times greater than the earth in volume or surface, so that many earths like ours could be lost in its bulk and nearly ten of them could be set in line on its diameter, it would take but three hundred and ten earths to balance its weight. This is because its material is so expanded by heat.

The spectroscope has shown us the elementary material of which that planet-source, the sun, is composed. This stuff is the same for all the planets, as all were cast-off portions of sun material. A certain amount would be of equal weight in any planet, but the size of that amount of material would depend upon how much it was expanded by heat.

The great volume of Jupiter affords a good field for telescopic study. Upon its surface a remarkable series of dark lines, called belts, have been found. These pass around Jupiter, parallel to the equator. Various pictures of the belts have been made. They all differ, because the belts are ever changing, in width, outline, and distance from each other.

Jupiter has spots which change less than the belts. One, called the "Great Red Spot," has been visible for twenty years. There are also white and black spots on the surface.

What are the belts and spots of Jupiter?

It is probable that the belts are layers of cloud, parts of which reflect the sun's light more perfectly than the surface of the planet does. The spots are probably rents, or craters, on his less than solid surface. The storms produced by the planet's highly heated condition convulse these cloud belts of his atmosphere with storms. The heavy cloud belts about Jupiter suggest an atmosphere loaded with moisture. When this moisture is high enough above the burning surface of the planet to escape from his diffused heat, it would condense into rain and snow and hail, as does the moisture which rises from the surface of our earth by the action of solar heat. Such condensed vapor would speedily be reconverted to steam when, descending, it again came within the influence of the planet's heat. In the intense heat about the vast glowing surface, atmospheric excitements, such as we call cyclones, tornadoes, and whirlwinds, would be of constant occurrence, and of a fury to which the wildest storms of our earth would afford no comparison.

The heat of Jupiter, great as it is, is incomparably less than sun-heat, and while the sun in its glowing state gives off light in abundance, and heat of which some degree reaches to the most distant planets of his system, Jupiter has not sufficient glow to diffuse light,

nor sufficient heat to radiate heat beyond the close
envelope of its cloud atmosphere. Jupiter does not
illuminate his own little moons; they, like their pro-
genitor, enjoy the largess of their grandfather sun.
A noted astronomer says : "Enough has been
demonstrated to enable us to pronounce on the
question as to whether Jupiter can be a body inhabited by living beings, as we understand the term.
Obviously it cannot. The internal heat, and the
fearful tempests seem to preclude the possibility of
organic life, even were there not other arguments
against it. It may, however, be contended, with perhaps some plausibility, that Jupiter has in the distant future the prospect of a glorious career as a
residence of organic life. The time will assuredly
come when the internal heat must subside, when the
clouds will gradually condense into oceans. On the
surface then may dry land appear, and thus Jupiter
may be rendered habitable."

It naturally occurs to us to ask, "Why are Neptune, Uranus, Saturn, Jupiter in such a state of
furious heat, a condition of material scarcely so condensed as fluid, when they were cast off by the sun
into space, so long before the earth, which has been
the home of organic life for myriads of centuries?"
The answer is found in the fact of the immense size
of these bodies and the rapidity of their motion.

The earth, so much smaller, contracted and cooled in
incalculably less time than these planets have demanded for similar processes; and numberless more
ages must pass before the solid crust, the ocean
reservoirs, the soil prepared by heat, pressure and
disintegration, can be productive of vegetable life
upon these elder planets.

Our moon, so much smaller than our earth, has, in
its briefer life period, had time to lose all its original
fire, and become cold and silent. The condition of
the moon hints to what state, in the long succession
of ages, our earth may be journeying; this case of the
earth and the moon is an object lesson of the case of
the planets, sent off before earth into space, but still
in the physical condition of the earth shortly after
its severance from the mass of the sun. As far as
material is concerned one can see no reason why the
life history of these planets should not correspond
with our own.

Modern discoveries have shown us that the elementary substances present in the other bodies of the
universe are those which form our globe, and we
must deal with the questions arising about the other
planets on the ground of similar material. We can say,
with some assurance, that heat, motion, contraction,
the forces which have built our earth to what it is,
will do the same work on other planets—on Jupiter.

When we meet other questions, as " Why, when Jupiter is twelve hundred times larger than our earth, is he only three hundred and ten times as heavy?" we cannot reply, " Oh, he is made of lighter elements." We must say the elements are the same, but in a different condition of heat and expansion. Such are some of the interesting directions of study suggested by the clear white splendor of Jupiter as the unaided eye marks his journey along the skies, or when through the telescope we detect about him his beautiful system of five small, clear, white moons.

These moons of Jupiter have lately aroused much discussion. There is, in South America, at Arequipa, an observatory which enjoys superior advantages for astronomical work, on account of the great purity and clearness of the atmosphere on those Andean heights. At this Arequipa observatory certain singular phenomena have been noticed in these little moons of Jupiter, phenomena known nowhere else in the system.

Our moon is, as says Dr. Young, "a solid globe of rock." The moons of Jupiter seem to be clouds of fiery mist, or fog, or whirls of dust and meteoric stones. In fact, they seem very closely to resemble in matter and state of excitement the planet from which they sprung, and if their condition is as suggested by Arequipa observers, then they are satellites

in their very earliest stages, only recently separated
from the parent globe. As we shall more fully understand when we come to discuss tides, there is on
all planets tidal action, and this tidal action on the
vast planet Jupiter, which is in such a condition of
heat and tenuity, powerfully affects its little incoherent moons. Instead of being a globe steadily revolving on its axis and sweeping along a regular track,
each little moon of Jupiter twists, writhes, and dis-
torts itself, much in the fashion of some sea-anem-
ones. These moons change their forms from round
to oval, and back again continuously, so that their
real shapes are hard to determine. The first moon
is now said to be shaped like a lemon, and to bowl
along its track, turning end over end, and this not in
the line of its orbital motion, but reversing it. The
second moon is flattened, as if a lemon had been
partly squeezed, or "like a cake of toilet soap," as
one observer says. The third is the largest of the
set, and is a miniature of the parent planet, orange
shaped, but instead of revolving on its shortest axis,
like a reasonable well-bred little moon, it whirls
round and round as if twisting on a string, and in
such a way as to keep the same face, or nearly the
same, always to the planet. The fourth moon varies
these antics by keeping an edge toward Jupiter, and
is of a much darker complexion than the others.

Now, the variations of these moons from the globe shape are, while definite, not very pronounced, and seem to be due to the incoherent state of the moon material, and the very powerful force of attraction exercised over them by the great Jupiter.

As Mars recedes from the eastern sky, Jupiter takes his place and would challenge the admiration and close observation of every star-lover by his singular beauty, even if all these marvels of his size and system were still as unknown to us as to his ancient beholders. Sirius is perhaps the only rival of Jupiter in brightness beheld by the naked eye ; but Sirius is so far, far off, and Jupiter is our brother planet, an elder and greater brother of our little green earth, a hundred thousand times greater than the earth, rivaled in size only by the parent sun. Jupiter is the privileged brother, the Judah of the sun family.

CHAPTER VIII

THE SUN'S SMALL CHILDREN

"The skies are painted with unnumbered sparks,
  They are all fire, and every one doth shine ;
  But there's but one in all doth hold his place."

WHAT planet comes next after Jupiter? we ask
when we settle ourselves for another thought journey
to the starry skies. Mars comes next. Between
Mars and Jupiter there is a vast space, which we
must visit before reaching Mars. We understand
that there are no stars within the bounds of our solar
system. Stars are suns like our sun, and each has
its own station and system. The very nearest to us
lies many billions of miles beyond Neptune. Between the orbit of Neptune and the sun is the
dominion of the sun, and inter-planetary space is
dark as inter-stellar space.

The distance from us of all the heavenly bodies is
so great that to the eye, or to the telescope, the re-
moteness of stars or planets appears equal. We need
to measure to ascertain true distances. This far-offness has caused planets and the moons of planets to
be mistaken for stars, until closely investigated.

Is there then a great empty space between Mars and Jupiter?

It was formerly supposed to be empty. Various interesting discoveries have, however, been made within its limits. Columbus, we know, decided that there must be land which might be found by traveling westward. He sailed away and discovered America. Astronomers having reasoned that there must be planets between Jupiter and Mars, traveled thither by telescope, and discovered new spheres.

It had long been thought that the emptiness of so large a tract was not in harmony with the construction of the rest of the solar system.

An astronomer named Bode found that a remarkable law governed the relative distances of all known planets. The space between Mars and Jupiter seemed to offer the only exception. Possibly then this might not be empty space, but contained planets that agreed with the general rule.

"If there is any planet between Mars and Jupiter," the astronomers said, "it must be very small, for being so comparatively near, a large body would have appeared to the unaided eyes."

About one hundred years ago, astronomers in all parts of the world arranged a plan of search.

An astronomer named Piazzi, in Palermo, Sicily, undertook observations on a plan of his own. He divided the stars into groups of fifty each, and observed and noted the places of these for four successive nights for each group. The air of Sicily is peculiarly fitted for astronomical observations; it is so clear and still.

On the night of January 1, 1800, Piazzi noted what seemed to be a small star in the constellation of Taurus, or the Bull. He marked its place on his map, No. 13 of the group in hand. For three succeeding nights he marked the position of this star, and then compared his observations. He found that No. 13 had changed its place every night, and was prancing about like a little boy just let loose from school. In fact No. 13 was not a star at all, but the sought-for planet between Mars and Jupiter!

The attention of astronomers was now directed to this planet. It was found to travel about the sun in an orbit of its own, not acting as the satellite of any other planet. It was named Ceres, after the goddess of harvests, a goddess of old supposed to inhabit Sicily.

When in its orbit little Ceres passed out of sight, astronomers feared that they should lose it entirely, and not be able again to fix its place. A young German scholar named Gauss, here made a great contribution to science. He took the three places in which the little planet had been seen, and drew for

6

Ceres an ecliptic, or planet-track, of which the sun was one focus, the track passing through these three known points. The ellipse being mathematically correct, having discovered from the three places the rate of travel of Ceres, Gauss could point out whereabouts the planet should be at any given time.

This seems very simple when it is explained. It is the first time that counts in all such investigations; and young Gauss acquired great fame and estimation for his work.

Was it likely that little Ceres occupied alone so great a place? All astronomers were now endeavoring to find similar small planets. During seven years, three more, Vesta, Juno, and Pallas were discovered, and the number was supposed to be complete. Forty years passed, and then, in the same field of space, scores of other little planets were discovered, one by one, by different astronomers. These were named, weighed, measured, and their orbits traced out, until by April, 1891, three hundred and forty had been found. There may be hundreds more, for the place they seem to occupy is very great.

What are these little bodies? How can we account for their place and number, for the eccentricity, as it is called, of their paths? This means that they are irregular in their orbits, departing very markedly from the circular.

Some astronomers suggested that long ago there might have been one great planet in this orbit, and that it had exploded from the violent action of its own gases, and its parts had been scattered into fragments. These fragments having gravitated about many centres of attraction, had formed the little planets which are called, collectively, the asteroids.

Another suggestion was, that a great planet having been cast off by the sun next after Jupiter, some errant comet had come that way, and the planet had been destroyed by a collision, and not by an explosion. Further and more careful study makes it fairly certain that the asteroids, or little planets, were formed just as their greater brothers were, that had been cast off by the sun in a quick succession of small rings—a certain high fashion of fire-works!

The asteroids lie about eighty millions of miles beyond the planet Mars. They vary in size; some are only a few miles in diameter, others are a great many miles. Being so much smaller than the other planets of our system, they must have been among the first to cool. They are probably even more solid than our earth, for our earth has a liquid or gaseous centre, or core. The tiny asteroids, most likely, have cooled and hardened entirely through.

Upon their little spheres may be continents, oceans, rivers, mountains. They might be very interesting objects to study. Unfortunately they are so small that the best telescopes cannot bring into view features upon the disc of any one of them. None of them are visible to the unaided eyes.

Owing to their small mass they are easily drawn and swayed by the attraction of other bodies, which may cause the deviations and eccentricities of their paths. We can also reason concerning them that from their minute size, whatever atmosphere they have must be very light and thin, rare, as it is called. They are of great value in astronomy, as aiding in calculating the place, motions, distances of other heavenly bodies.

If we were upon one of the little asteroids we could travel over its entire surface in a short time. I say "if," for there is a large "if" in the way—if there were any atmosphere that one could breathe; if there were any water for one to drink; if the attraction of gravity on so small a body were strong enough to keep one's feet upon the surface, and prevent one from tumbling off into space.

Take an asteroid eight miles in diameter, our earth being about eight thousand. A million such asteroids would be needed to make an earth like ours, if the solidity were equal. If the mass is one thousand times less, attractive force would be less in proportion. If here we can throw a ball up twenty yards,

there we could throw it twenty thousand yards. If
here we can jump up one yard, there, we could go
up one thousand. We and our ball might get off
too far ever to be drawn back.

Do we think that very amusing? It might have
terrible disadvantages. One might wish to jump
over a house or to a tree-top, and might inadver-
tently give such a send-off of tangential motion that
one might never get back. Here we jump up and
are drawn back, or we fall and are drawn down by
the attractive force of the earth pulling us and all
things within its reach, toward its centre. The attractive force on a little asteroid is so small that the
tangential force exerted by a human boy might free
him from the planet and set him loose in space!

So interesting is the search for these little planets,
and so curious and attractive are the questions concerning them, that almost all astronomers give themselves more or less to the search for them. "Asteroid
hunters" is a common expression nowadays. Since
it was finally decided that—

> "No suns had clashed, no planets burst;
>   The worlds whirl on their way;
>   The day makes beautiful the night;
>   The night makes glad the day—"

the asteroids, as little individuals built on the same

plan as their greater brothers, are of increasing attractiveness.

Until 1892 the search after new asteroids was slow and laborious. The method was this : A portion of the blue arch was mapped out ; all stars therein seen were carefully set down, and their places noted. This done, when such a field came again opposite the sun, the telescope swept it carefully, minutely examining every body visible in it, to see if some glittering orb appeared which was not noted on the chart. If such an interloper was found, still more careful study of it must be made, to be sure that it was not a variable star, or a star overlooked in a previous search. A few days, perhaps even a few hours, would settle these questions. Then a new query rose, if this new object proved to be really an asteroid, was it an asteroid now for the first time picked up by the telescope, or was it one of the older discoveries? For there are several of the first found asteroids, which are what is called " adrift," that is, in their flight about the sun they have for a number of years escaped observation.

One great effort of our present age seems to be to minimize labor, and so the labor of finding asteroids has been greatly lessened. Human ingenuity has again made a triumph. Asteroids are now hunted after with a camera! This camera has a peculiar

lens, six or eight inches in diameter.   The camera is
sometimes strapped upon the tube of a telescope, for
convenience in following star motion.

The observer then photographs the region where
he is searching for asteroids.   Photography is not
careless, it never overlooks a star.   It is a minute
truth-teller.   Points which would escape the human
eye gazing through a telescope are rigorously set
down by the camera.   The camera also will cover a
much larger field of observation than the ordinary
telescope.

How large do these thousands of stars appear
as seen through the tell-tale camera?   Each
one on the negative will be a little round dot.
And here is a fact worth noting.   While the stars
are each represented by a distinct dot, a planet
on the other hand, is not a dot, but drawn
out into a little streak, because of the appreciable
motion of the planet during an exposure of the
camera of some hours.   Now the asteroid being a
planet, and moving with velocity, prints itself on
the camera as a little streak, and so the plate has
been known to catch two or three of these tiny
travelers at a single exposure.   In the year 1893
forty asteroids were discovered in this way.   This
simple and ingenious fashion of search was first em-
ployed by an astronomer named Wolf, in Heidel-

berg, Germany. Professor Charlois, at Nice, used the same method. Four of the forty detected in 1893 were old strays, asteroids once seen, known, named, and then escaped. Also fourteen of these forty, while probably new asteroids, need further examination before they can be fully determined upon. Some of these new-found asteroids are from fifteen to twenty miles in diameter.

One astronomer, writing of asteroids, says he is "seriously afraid that several thousands more may come into view, while their present number is already so large as to be embarrassing." Who, he inquires, is to spend time looking after all these little strangers, and making a chronicle of their doings? The multitudinous babes of the sun family are then a troublesome element, though no sound emanates from that vast sky-nursery which lies between Jupiter and red Mars.

CHAPTER IX

RED MARS

" There is no light in earth or heaven
But the cold light of stars,
And the first watch of night is given
To the red planet Mars."

LET us see if we can find Mars for a study; it is a
red star, and always shines very brilliantly, but the
best time of all for seeing it is when it is in opposition.

What do we mean by that? When is a planet in
opposition? "A planet is said to be in opposition
when it is on the side of the earth opposite the sun.
The earth is then between the sun and the planet;
the planet receives the full sunlight, and we have the
advantage of looking away from the sun while we
observe the planet. When Mars is in opposition it
is nearer the earth than at any other part of its
orbit. Mars is one of our nearest neighbors in the
skies, and we know more of him than of any other
planet, except the moon and the earth. When we
have found three clearly red stars, how can we know
which of them is Mars, and what are the other two?

One is Aldebaran in the constellation of Taurus; one is Beltegeuze in Orion. Let us get the planetarium and set it with its face to the north. Note where Mars is set on the globe of the planetarium, and then follow with your eyes the line of Mars up to the sky.

Now we have it! It is the brightest of the three.

For some time we will be able to watch Mars each evening, but there are parts of the year when Mars rises and sets with the sun and is lost in his light. If we were spirits roaming through the skies we should see two planets much alike in shape and general appearance. Perhaps at first we could not tell them apart. Then we might learn that one of them was called Earth, was twice as great in diameter as the other, and had but one moon. The smaller of the two is Mars. If there are people on the planet Venus, who look at us through telescopes, we appear to them much as Mars does to us.

Looking at Mars we see a globe, slightly flattened at the poles, revolving obliquely on its axis, and turning over once in twenty-four hours. Its journey around the sun requires one year and almost eleven months. Thus the seasons of Mars are double the length of ours, while the day is of the same length that we have.

Mars is but little more than half the size of our

earth. Until recently is was supposed to have no moons. It was called "moonless Mars," "lonely Mars." In 1877 Mars was in opposition, and at the place in its orbit nearest the earth ; the opportunity for observing it was exceedingly fine. Professor Hall, of the Observatory in Washington, discovered two moons revolving about Mars. One is eighteen miles in diameter, the other twenty-two. The outer moon requires thirty-one days for its trip about Mars, but the inner moon traverses its little circle in seven and a half hours. This moon rises and sets three times each day of Mars ! The planet is not particularly benefited by such an active little moon, because it is so near that much of the time it is hidden in its course.

The behavior of the inner moon of Mars has no parallel in the solar system. As moons gradually drift away from their planets, being nearest at their first formation, the inner moon of Mars may be of comparatively recent origin, and may, in time, make a wider circuit.

"What are the names of Mars' merry little moons?" Professor Hall named them after the two horses which Homer tells us drew the chariot of Mars, the god of war, Dismay and Rout; in Greek, Deimos and Phobos. More than a hundred years before these moons were discovered, Dean Swift, in his "Gulliver's Travels," stated that the star-gazers on the Flying Island had discovered two moons for Mars, and that one of them traversed its orbit in ten hours. This was truly very close guessing.

From Mars we learn the weight of our earth. If Mars had no attraction to sway him but the sun, his path would be forever the same. Our earth, his nearest neighbor, is large enough to disturb his orbit by attraction. Mars is pulled toward the earth. Astronomers take the position in which Mars would be if attracted only by the sun, and then take the place into which he is gently pulled by our earth. The difference between these two is all due to the pull of the earth. It is then necessary to calculate how large a mass is required to exert so much attractive force; thus the mass or weight of our earth is learned.

When Mars comes nearest the earth, features on its disc are clearly observed through the telescope, and Mars has been well studied. The poles of the planet are found to be capped with snow, like our own. Maps have been made of the surface of Mars, and these look curiously like a sketch map of our world, drawn on "Mercator's Projection." There are singular markings on Mars which can scarcely be accounted for except by the presence on the planet of bodies of water.

The presence of water is also suggested by clouds in the atmosphere of Mars. If the snow-caps melt we should expect clouds from the evaporated moisture. The polar snow-caps increase and diminish, as might be expected, during the long summer and winter seasons of the planet. We have no means of knowing whether the air of Mars, in composition and density, would be suited for breathing by beings like ourselves.

Owing to the small size of Mars, it has no doubt cooled nearly or quite through, but still the density of the planet is less than that of our earth.

When speaking of the asteroids, we spoke of the force of gravitation as dependent upon the mass of the body in question. As Mars is but half the size of our earth, its attraction of gravitation is but half as strong as we know here. Anything weighing a pound here on the earth's surface would weigh but half a pound in Mars. If here one can lift a ball weighing fifty pounds, there, with equal ease, one could lift a ball weighing one hundred pounds.

Walking, jumping, any activity would be but half as fatiguing upon Mars as here. On the other hand, the circulation of the blood, the action of heart and lungs would be so violently increased that beings constructed as we are might be unable to exist in such circumstances.

As our earth and its inhabitants have been so carefully adapted to each other, we may conclude that if there are inhabitants in Mars equal harmony exists between them and their abode.

Either from its red light or its name, the ancients thought Mars a very cruel and dangerous planet, the cause of nearly all of the disease, war, famine, and misery upon earth. Mars and Saturn were regarded as a pair of twin demons, full of evil influences.

Why is Mars so alarmingly red?

Some have suggested that the vegetation of Mars is red and perennial; others fancy that the earth and rocks are of a deep red. The " water lines " on Mars are bluish; the polar caps are white. Thus Mars is a kind of American-flag planet—red, white, and blue.

The year 1892 was very favorable for the study of Mars, as the planet was nearer us than it will be again before 1909. Sixty-eight drawings of the disc were made. From these maps we find that Mars has more land than water. Also on the surface were found some remarkable straight lines, which seem to be water, and are known as the "canals of Mars."

As Mars is further from the sun than our earth, and is no doubt cooled quite, or nearly quite through, we might expect it to be colder than our earth. As far as we can discover from telescopic observation, however, the climate is much the same as here.

The state of the ice caps on the Martian poles gives some indication of the temperature. We find that these shrink, by melting, during the long summers. We also find that the equatorial regions of Mars seem to be free of snows, as are the tropics of this world.

On the whole, Mars has proved one of the most interesting and useful objects for astronomic studies, and, as telescopes improve and observers increase, we are likely to know more and more about it. The nearness of Mars has given rise to many extravagant suggestions and romances. It has been said that in some way communication could be opened between Mars and the earth. Any inhabitants of Mars must have intellects, and any intellects would recognize a geometrical figure; therefore, if men built on some vast plain—as the Sahara or a South American pampas—a proposition from Euclid, the Martian-men would recognize it as an output of sense on the part of Earth-men and would reply in kind!

Another proposition has been that as Mars is much older, in all probability, than our world, it must have been much longer inhabited, and the Martian people have had time far to surpass us in all knowledge, discoveries, and inventions. Therefore being so much wiser, they will soon find some way not only of opening communication with our earth, but of getting here. In fact, the Martians must be eager to reach here, because owing to its greater age, smaller size, and longer distance from the sun, Mars must be not only too cold for comfort, but too cold to live upon at all, and its inhabitants must be anxiously looking about for a place to colonize. Many people in the United States have been complaining bitterly about the influx of emigrants from Europe and Asia; if, through the skies, emigrants from Mars begin to drop upon us like a storm of hail, or even like the large silent flakes of a snow shower, a wail would be lifted from all the world. Our only hope would be that our climate would be so much too hot for them that they would perish, unless they hastened their departure. Even the poets have indulged in absurdities about Mars. Mr. Kipling considers him as a horse wildly careering through the skies, and says:

> "Hanging like the reckless seraphim
> On the reins of the red-maned Mars."

Professor Holden thinks that if by "reckless seraphim" the poet means star-gazers, that he is quite right, and they have been more than reckless enough. Really scientific men are careful not to be hasty in statement, nor to indulge in theories at the expense of assured facts. Careful observations made at the Lick Observatory in 1895 raised a doubt whether there were as much water on Mars and in its atmosphere as had been hitherto supposed. Observations made by spectroscope at the observatory at Mount Hamilton show that Mars is amazingly like our moon, and if it should prove that Mars is in the condition of our moon, dry and cold all through, then verdureless and waterless, all question of " men on Mars " will be finally set at rest.

When Mars is in sight anxious amateurs and astronomers spend their time in observing it; when it is lost in its sun-bath, drowned in light, they spend time talking and writing about it; all this considered, it seems that we really ought to know more about Mars than we do! Professor Young considers that the Martian canals are real and that great changes in them accompany the waxing and waning of the polar ice-cap of Mars; he thinks also that, as compared with ours, the atmosphere of Mars is very rare.

Mr. Lowell, possibly because he was writing charming papers for a magazine, directed to a not critical public, theorized that Mars has no hills, no mountains, is a dead-level land like Egypt, and that the melted snow-caps send still floods over all its surface, carrying fertility with them. He holds that the dark

7

spaces are not seas but plains covered with dense
forests. He also thinks that the canals of Mars
must be artificial because they are so straight, and
that what we see is not the canal with its waters, but
the exuberant vegetation beside them. From the
size of these canals, which he claims must be artificial,
he asserts that the men who made them must be
gigantic ; and that these huge men must have a
strength a hundred times greater than the inhabitants
of our earth, and an infinitely greater knowledge of
mechanical appliances. On the whole, after hearing
all this, we feel very glad that so many millions of
miles lie between us and " the red planet Mars."

Professor Young calmly remarks that against all
these theories of Mr. Lowell stands the fundamental
doubt whether so small and distant a planet can have
anywhere more than a life-destroying coldness of
temperature, and another doubt, namely, whether the
ice-caps are ice at all or some substance quite different. The observatory of Mr. Lowell is in Arizona,
and continuance of telescopic work there has convinced him that " Mars does not present the same
appearance for two successive seasons, and the
differences are not confined to details but to large
and prominent features " upon which at times
theories have been founded. Thus it seems that our
red neighbor in the skies is a very shifty and deceit-

ful planet, and no dependence can be placed upon
his revelations concerning himself. The fact is, that
we should remember that in all observations made of
the heavenly bodies the earth's far extended atmosphere must count for something, and lines might
seem to be upon the planet under observation, when
really they were to be attributed to some action or
condition of the earth's atmosphere, and it is only
many times repeated observations of the same regions
of any body that would correct errors which arise
from our own conditions.

CHAPTER X

"Brightest seraph, tell
In which of all these shining orbs
Hath man his seat?"

WHERE shall we make our journey to-night? we ask. We shall stay closely at home to-night. Our Earth is the planet next in order of formation as we travel sunwards from Neptune through the ranks of the solar family.

After that glowing ring which coalesced into Mars was cast off, we do not know how long the sun rested before a ring seven times as large went wheeling into space. When this new ring arrived at a distance of some ninety-one or two millions of miles from the sun, it was stayed by attractive force getting the better of tangential force. Turning over and over in the direction from west to east, which it had received from the sun, this ring came together into a ball, continued to make a revolution upon its axis, and to swing in a great orbit around the sun.

The orbit of the new planet was elliptical, like the orbits of the other planets. The concave side of the

100

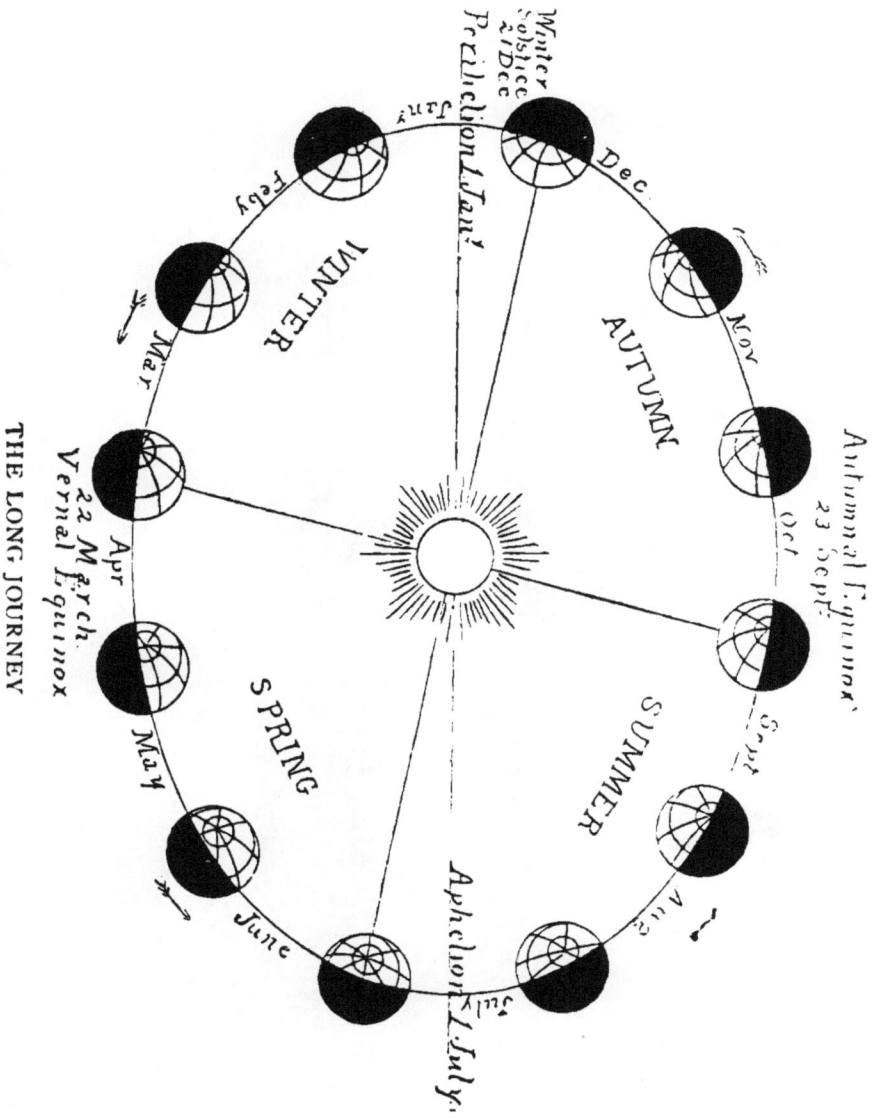

THE LONG JOURNEY

ellipse is always toward the sun. The rotation of the earth on its axis occupies twenty-four hours; the synodic, or yearly journey is a little over three hundred and sixty-five days. If the plane of the earth's motion were not inclined, or, as we might say, tilted, to the plane of the ecliptic or pathway, the days and nights would be of exactly the same length the year around. This inclination, together with the time of its trip through its elliptical orbit, produces the seasons. We have four seasons, each of three months duration.

If the earth turned over and over, but did not travel about the sun, there would be only one season. Such an "if" could not occur, because if a planet did not travel along an orbit about the sun, as soon as tangential force was exhausted attractive force would be so strong that the planet must tumble back into its source. But for the restraint exercised by the swift roll in an orbit the moon would have tumbled back into the earth, the earth and the other planets into the sun again. So the sun would have been like the fabled monster-god Saturn, who devoured his children.

The turning of the earth on its axis causes one side to be presented to the sun, while the other side is in shadow; we call that shadow night. With the turning of the earth, that line of shadow moves slowly along the surface, night passing into day as it moves. When, owing to its inclination, or tilting on its axis, more of the earth's surface is exposed to the sun, the days of that exposed portion are longer; when a part is tipped away from the sun the days in that portion are shorter. It is this tipping, or inclination, which causes the sun to seem to be in the south sky during winter, and to the north in the summer.

It occurs to us to ask why the earth moves from west to east, when it seems to move from east to west; that is, our earth seems to us to stand still, and the sun to move, and the sun appears to move from east to west. We have all been on the steam cars moving out of a depot. As the train started quietly, it seemed to us that the train next us was moving, and our train was standing still. Or, when our train stood still, and the train next us started, it seemed as if our train was in motion and the other at rest. As we are on that moving planet, the earth, it seems to us to stand still, while the sun, which, so far as we are concerned is still, seems to be journeying swiftly about us, as the ancients supposed to be really the case. Again, when we have traveled in the cars, trees and houses seemed to us to be rushing, not with the train, but back toward the point of starting. Thus, as our earth-motion is swiftly from west to east, it causes the apparent motion of the sun from east to west.

Having settled this point, we need to know more about night and day.

Here, in the first place, is a little experiment which we have arranged for illustration. On a rubber ball are painted two white poles, a red equatorial region, and two green temperate zones. We have run a long hat-pin through the axis of the ball, and hold it tipped, as the earth is tipped, to the plane of the ecliptic. Now, here is a long cone of paper, which fits upon the equator of the ball. We pin this gilt star to some fixed place to represent the sun. The cone represents the shadow cast by the earth. The exposed part of the ball lies fair to the sun, and represents day. Hold the cone steadily upon the side of the ball opposite the sun, that represents night; the opaque earth shuts off the sunshine from that half of itself. Now, slowly revolve the ball upon the pin; some of the part that has been in the cone-shadow creeps out, and equally some that has been exposed creeps in. The side which was in the light at first had a blue 1 upon it, and the side within the cone a yellow 1. Now we see the blue 1 is in the night-cone, and the yellow 1 has crept into daylight. The days of our northern half of the world are at their longest at mid-June, and are shortest at mid-

December. At the equator their length is more nearly equal, and they do not have the long twilight at dawn and sunset that we have.

Like the other planets, our earth is round, bulging at the equator, and slightly flattened at the poles. All heavenly bodies but comets have this round shape. The sun, the moon, the discs of the planets, exhibit this rotundity. Comets seem to have roundish heads, from which stream back tails or trains of burning vapor.

The ancients supposed the earth to be a flat, circular plain. Several simple illustrations prove to us its rotundity so clearly that we wonder that it was ever doubted.

From some balcony we have watched the ships putting out to sea. If they disappeared on a level plain, merely because they went out of the reach of our vision, a ship would, as a whole, grow smaller and smaller, and so fade from sight. Instead of this the hull sinks out of sight first, while the tall masts remain in view for some time longer. As says Coleridge :

> " The ship was cheered, the harbor cleared,
>   Merrily did we drop
>   Below the kirk, below the hill
>   Below the lighthouse top."

Fancy that a pin stuck in our rubber ball represents ourselves, and a little peanut shell with a splinter for a mast is a ship sailing away. We keep our eyes on a line with the pin's head. As the shell-ship moves down the roundness of the ball we lose sight of it, but the little mast is in view somewhat longer. When steamers come toward us along the sea we see the smoke from the funnel first, then the funnel, and then the hull.

Another proof of the shape of our earth is that its shadow cast on the moon in an eclipse is round, and like the shadows cast by other planets which we know to be spheres. We can also prove the rotundity of our earth by measurements which are too difficult and abstruse to suit our present purpose.

The surface of this globe, the earth, is composed of land and water, about one-fourth land and three-fourths water. The diameter of the earth is nearly eight thousand miles, and its circumference twenty-four thousand. It has polar caps of ice and snow.

Early in its history our earth cast off a gaseous ring, which became a globe and travels around the earth, as its moon or satellite, at a distance of two hundred and forty thousand miles. The difference in size between the earth and the sun has been likened to a grain of mustard-seed revolving about a cocoanut. If the sun were a hollow globe and the earth were set in its centre, with the moon whirling about it two hundred and forty thousand miles away, there would be very nearly as much distance still left between the moon and the shell of the sun as there is between the earth and the moon.

The crust of the earth is solid, being formed of rocks and what we call earth, or mold, made of ground-up rocks and decayed vegetable matter. The rocks and sand form vast basins wherein are bodies of water called seas and oceans. Evaporation from the surface of the earth condenses into clouds, which are always in our atmosphere. Our earth has more clouds than Mars, but not nearly so many as Jupiter.

While the crust of the earth is solid it is generally supposed that within that crust are several layers of matter of different densities about a gaseous core.

The mass and density of the earth have been many times measured. During 1893 Professor Berget in France made a new measurement, which was distinguished from others by the freshness and beauty of its method. Berget borrowed a lake from its owner, Monsieur Curel. The lake was artificial, was eighty acres in extent, and had water-ways and gates by which its level could be rapidly changed. This lake was at Habay-la-Neuve. Berget had an instrument for measuring the force of gravity. This instrument has a column of mercury "balanced against the elasticity of a certain confined body of hydrogen," both shut up in a vacuum, and kept in a constant temperature. Now, just as in a thermometer or barometer, a change in the temperature or moisture of the air causes a rising or falling of the mercury inclosed in a tube, in this instrument the mercury is moved by the least change in gravity, and the gravimeter, or gravity measure, is capable of registering so small a motion as the one hundred millionth of an inch.

Professor Berget, by means of the water-ways and gates, lowered the level of the lake somewhat over a yard—one French meter—and then examined the change in gravity registered by his instrument to see what was the value of that amount of water in gravity or pulling force. Measurements in gravity and density taken before were not changed by this new method and its results. The interest lay in the curious delicacy of the instruments used and the originality and simplicity of the plan.

"I don't see," said a thoughtless person one day, "why any one cares about the density of the earth or the force of gravity. What does any one want to know about it for? What difference does gravity make to me? Suppose the force of gravity should change all in a minute, who would care? Who would know it?"

There is a deal of absurdity in such talk, but
we answer: Every one would know it, could not
help knowing it, and would care a good deal. It
is a fortunate affair that the laws of the universe
are reliable, unchangeable, and there is no dodging
them, no shirking of their consequences. Let us
suppose that the attraction of gravitation could all
in an instant double. Then all power needed to
effect any result would need to be doubled. It
would require twice as much effort to lift the foot
from the ground to make a forward step. One might
well say then that "my feet felt as if made of
lead." One would soon grow weary at that rate.
The weight of every object would be doubled. If
one were buying ten pounds of sugar and had paid
for that amount one should receive for the money
what had hitherto passed for five pounds. The
wheels of the cars would rest so much more heavily
on the track that two engines would be needed to
do the work of one. If one fell, one would come
down with twofold violence, and so on.

Suppose, on the contrary, the pull of gravity suddenly lessened by one-half, then the barrel which
had held two hundred pounds of flour would hold
but one hundred by the new weighing; a student
devoting himself to athletics would be charmed to
find himself making a "standing high jump" that

beat all known records; a ball tossed high might sail out of sight and leave its small owner weeping; the axe brought down with customary stroke on a billet of wood would be endowed with so little force that it would make but a small cut. "The times would be out of joint" for us, sure enough, and all our calculations overthrown.

CHAPTER XI

"A boundless continent,
Dark, waste, and wild, under the frown of night,
Starless exposed."

"WHAT I want to know is," says some one, "why this force of attraction, pulling together a ring of matter, always makes a ball-shape."

Because the pull is even all around the common centre—the centre of gravity. Just as much material is pulled at one place as at every other. This, of course, causes roundness.

Yet the planet balls are not even or round, absolutely; they are flattened at the poles and bulged at the equator. We will try to illustrate, first as to the roundness, then as to the deformity. Gravitation causes all liquids to assume a round shape in falling. A drop of water is round. If you put a drop of thick oil into water, with which it cannot mingle, the drop of oil remains round, suspended in the water.

Let us consider how they make shot. High towers are built, called shot-towers. The melted metal is poured from the top of the shot-tower, and falling

110

to the ground it cools in its descent into round evenly-made shot. The size of the shot is regulated by the meshes of a sieve on the top of the tower, through which the molten metal is poured. The shot fall into a tank of water, where they complete their cooling. Shot towers are from one hundred and fifty to two hundred feet high.

This story of the shot shows us how liquid or gaseous or other soft bodies are pulled by gravitation into roundness. As to the bulging and the flattening, which deform the roundness, they come in this wise: The shot being so small, cool almost instantly, keeping their round shape; but planets, even the very smallest, are immense bodies of intensely heated gaseous matter, and cannot cool quickly. Each turns upon its axis, and whirls through its orbit in a state of incandescence. Continuing these double motions, "turning over" and "getting on," each sphere cools and solidifies slowly.

Imagine a ball of soft material set to whirling very simply; fancy you were spinning a ball of half cooled molasses candy upon a string. The consequence of the swift rotation would be a bulge in the circumference, a drawing in of the poles through which the string passed, and presently such an over-weighting at the ball's equator that some material would begin to fly off.

As there is only just so much material to account for, it is plain that where it swells at the equator it must shrink at the polar part, that is, farthest from the bulge. When the planet ceases to cast off material, and cools into permanent shape, the " deformity of the sphere," as it is called, that is, the equatorial enlarging, is a fixed feature.

How about the shape of the sun? That seems to be perfectly round, we say to ourselves; but do appearances deceive? It seems so, because of its enormous size; the sun has such equatorial enlargement, and polar depression as the planets exhibit. The extreme slowness of the sun's motion has prevented this deformity from being as marked as in some of the planets which rotate most swiftly. To on-lookers from distant nebulæ the rule of planet shape would seem uniform. How, we wonder, does our earth look to people on Venus—if there are any there—with telescopes?

Very much as Mars appears to us, except that it is larger and has a more clouded atmosphere. They see the polar snow-caps and the attendant moon. Some have suggested that just as the light of Mars is red, the light of our earth is a pale green, from our green vegetation, and our seas might shine like molten glass or Alpine glaciers. We wish one could tell us all about the way in which the earth has

changed from a ball of gas to what it is. Suppose
we could see it just like a panorama, a splendid
succession of scenes. We can trace a little of the
little that is known about it. We will remember,
meanwhile, that the changes which we describe are
probably those which sooner or later will overtake
all planets, and are now in progress upon them in
different degrees. We see great Neptune, Uranus,
Saturn, and Jupiter in their various states of heat
and fluidity, and we argue that our earth, as a planet
of like motions, shape, origin, and materials, has
passed through the same stages. We might reason
in another way also, that as our earth has now a core
of very hot material in a fluid or gaseous state, we
can infer the change from that hot core outward in
all its different stages.

We know that our earth has a heated core, because
in mines carried to a great depth the heat is almost
unendurable. Yet the deepest mines enter but a little
way into the earth's crust. Volcanoes, boiling springs,
or geysers are known in all parts of the earth, from the
polar circle to the tropics. Many of the islands and
mountain chains are of volcanic origin. Frozen
Iceland has violent volcanic eruptions. Its boiling
springs leap up from one to two hundred feet into
the air. These jets of boiling water, these floods of
red-hot lava, these veins of fiery ashes, all show that

8

the inner part of the earth from which they come must be in a molten condition.

The surface of the earth is cool to the touch, yet from it the interior heat is always imperceptibly oozing. We have, perhaps, visited a brick-yard; we know that a big brick furnace is built, in which they bake bricks and tiles. The thick walls of this furnace keep in the heat, so that it bakes the bricks red-hot through and through. Yet the kiln does not keep in all the heat. The outer wall feels warm to the hand. Heat is oozing from it, little by little, and the air carries it away.

In the same fashion from the surface of the earth the heat of its inner fires constantly passes, and each age the earth is cooler and cooler. We have no idea how many thousands of years this cooling process has been going on. The Creator never hurries His work.

The earth, as it has cooled and shrunk so much, must now be smaller than when it was cast off by the sun. It is much smaller in extent; but we need to remember that its weight has always been the same. Cooling and shrinking do not change weight. The weight of a piece of iron does not change, whether the iron is at its smallest size, perfectly cold, or is expanded by being red-hot, or white-hot, or still more expanded by being melted. The earth always weighed what it does now; therefore its attractive force was never less than now. Cooling, shrinking, and hardening have changed the state of the materials and the size, but the matter is always the same.

The earth constantly loses, but never gains any heat. The heat we receive from the sun radiates from the earth's crust, and never penetrates it very deeply. Think how soon the night shadow cools off the heat of the hottest day. Winter presently causes us to forget the heat of the hottest summer.

We know really very little of the interior of our earth. Great chasms or cracks in its surface, are, in comparison, no more than the little lines on the rind of an orange. Our deepest mines are, in comparison with the diameter of the globe, no more than the tiny dent you can make with a pin's head in the orange skin. In spite of all this, careful study of the small portion that is open to our inspection has enabled us to read the history of the whole.

When our earth first assumed spherical shape it was a huge ball of gas, probably two thousand times as large as it is now, all its materials being at their state of utmost expansion. This mighty globe of glowing vapor whirled and cooled, and the core, by pressure, became fluid. The outer portion acquired what is called a photosphere, or visible shining surface. This may be the present state of Neptune.

After a great lapse of time, by constant cooling the photosphere darkened, and a more or less solid crust was formed. The next state was probably that in which we now find Jupiter, the atmosphere violently stormy and full of clouds. After that would come a period of water forming and depositing, until the whole surface of the globe was under water. Do we ask, " Why are Neptune and Jupiter and other planets probably in a state that our earth passed long ago ?" They were probably formed a great while before the earth, not only that, but they are enormously larger. Cooling and hardening would require many more ages in such immense masses of material. After the water envelope of our earth was formed, we suppose that the crust under the water continued to cool and to thicken, then cracked and broke by shrinkage and the force of interior gases, steam, and molten matter. Being so rent, the crust was lifted, forced, tilted up by action of the interior elements. It rose above the waters in peaks, ridges, levels, and tilted sections. This lifted rocky crust became the foundation of future continents.

After many more ages the action of water and storms had corroded and ground up portions of rock into clay, sand, loam, marl, all that we now call earth. Thus the surface of the rocky continents was covered more or less deeply with material fit for vegetation, while the waters in the ocean beds had cooled sufficiently to become the abode of animal and vegetable life. Such life was in the water long before it appeared upon the land. The formation of land was very slow, accompanied by rising and settling of tracts of earth crust. Some of the substances just mentioned as composing soil were remains of generations of animal and vegetable life.

When this stage of fitness for life was reached the story of planet-building was nearly all told for our earth. It is a question often asked : " How long was this process of earth-building? How old is our earth?" Perhaps no question was ever asked that has received such widely differing answers, many of those answers being given as veritable scientific statement and demonstrated by mathematical calculations, based on certainties of physics. The age of the earth may be said to be written in its geological strata, but it is by no means written so plainly that he who runs may read it. The wisest differ greatly in their estimate of how long it takes for certain changes to occur. How long will it be before heat and pressure turn sand into sandstone? How many ages will be required for the waters to deposit in a fine rain of particles upon the floor of the ocean the comminuted shelly bodies of foraminifera until the great beds of limestone rocks were ready as a cheese for pressing? Some savants affirmed one period of time as needed for such radical changes in the state of affairs, and some another.

But long before shells, or sand, or other such now known material was, long ages were required for the ball of blazing, glowing gas to cool, contract, become semi-fluid, liquid, solid. Proceeding on calculations of temperature and the time required to change to a solid state from complete fusion, other wise men demanded say a hundred million of years. Now, when so many years as a hundred million are in question, a matter of forty or fifty, or even a hundred million more is but a trifle, so some concluded that three hundred millions of years would not be time too long for so vast changes, and others, equally as wise, asserted that from fifteen to thirty millions of years would give quite time enough.

"It is forty-five millions of years," says one, "since the globe was cooled and framed into a possible home for some forms of life; the first fossils date from forty-five millions of years ago." Thus the argument goes on, results forever changing, all that remains assured being that very long time—time beyond any possibility of real comprehension—

was needed in world-building. Are, then, these calculations and discussions wasted? By no means. Physics has come to the aid of geology, curbing extravagance of statement. It has been seen that accurate data have not yet, and, perhaps, never can be obtained, but that from widely differing premises conclusions not widely divergent may be reached, and in this searching after data for time statements many valuable discoveries have been made and the sum of human knowledge greatly increased.

CHAPTER XII

"And treat her as a ball, that one might pass
From one hand to the other."

HERE is a great, fair, full moon, all ready to be
talked about, a "hunter's moon" hanging full and
low ; a red harvest moon ; or a round, steadfast, clear,
cool, summer moon ! How near she looks ! It
seems as if one could shoot an arrow to her. Yet
we are told that she is two hundred and forty
thousand miles away. It is hard to realize or believe
in these distances. But then this distance is so short
in comparison with the space that divides us from
other heavenly bodies that we call the moon a near
neighbor, and large telescopes have made her surface
very well known to us. We know her as well as we
know some parts of our own globe. For instance:
What are those dark lines, like mountains ? They
are mountains, great volcanic peaks and ranges.
The whole moon-surface is a series of burnt-out
craters. The nearest likeness which we can find to
moon scenery is the upper portion of Mount Vesuvius. After you have passed the limit of vineyards
120

and cottages, you come upon a great broken slope of loose lava, ashes, scoriæ, and large lumps of volcanic rock. Not a blade of grass is seen in that silent, dry desert. Your feet slip in the loose material, old embers of dead fires. You come to the ragged rim of the crater, and look down into the great rough, broken bowl.

You cannot descend into Vesuvius, cross the crater, and come up on the other side of the rim, for the volcano is too frequently active, and the crust within the bowl is hot and thin.

The moon volcanoes were long ago burnt out, and if one could walk about the moon, one would find no hot lava. In fact, a traveler in the moon must be going up and down the crater sides, and across the lava fields all the time, for that is all the landscape there is. Looking at the moon we ask: "Has the moon had all the changes through which the earth has passed, from gas to fluid, then to solid?"

Most of them probably. Whether it ever reached a stage of habitable surface, such as our earth seems to be resting in now, we cannot tell. Perhaps those tall moon volcanoes which are like hollow cones with the tops cut off, were once in the flaming surface of the moon just such spots as we see now in the sun, and marvel about. The whole surface, so far as we know it, seems to be volcanic. Is the moon of any use, but to give us beautiful, light nights? What effect has this truant child on the mother orb? It causes the tides and the tides are of immense value to commerce and navigation. Many of our greatest seaport cities could not be approached by ships of any large size were it not for the help of the tides, each day giving deep water in the harbors. Many fishing fleets depend on the daily tides for their going and coming. Many of the great cities also depend on the tides to flush and clean out their sewer systems.

Some people think that the moon has much to do with the weather, and others talk of its influence on the growth of vegetables, and of " planting in the right time of the moon," but that is all a mistake, the moon has no influence at all in these directions.

Although the moon appears so large, quite as large as the disc of the sun, it is really the smallest visible object in the skies. Its nearness makes it seem large. It would take fifty moons to make a globe of the size of this world, and fifty million moons to make a sun. As to weight, the material of the moon is lighter than that of our earth, and eighty moons would be needed to turn a scale against our globe.

The moon travels around the earth in its orbit in twenty-seven days, and what we may think very strange is, turns over on its axis but once

A NEAR VIEW

in twenty-seven days also! The sun's rotation on its axis is accounted slow, taking twenty-five days, but the moon demands about twenty-seven days and a quarter for one revolution.

In consequence of the rotation of the moon being made in the same time as its orbital journey, we never see but one face of the moon. The other half is never toward the earth. There has been a deal of speculation as to what that unseen half of the moon is like. There are periods in the orbital journey when a little rim of the moon beyond the face constantly turned to us is to be seen on one side or the other, and as that portion shows the same volcanic desolation as the side we know so well, we conclude that the entire moon body is in the same state of barrenness.

We do not know whether the moon has cooled all the way through, or whether it has a core of soft lava. At all events, volcanic action has ceased in the moon for long ages, and no doubt the hardened crust is too thick to permit any eruption, even if there are interior fires.

Long ago the moon was much nearer the earth than now. It receded gradually, driven by tidal force, as if the earth in its swift revolving, having cast off the moon by tangential force, had slowly reached out a hand and pushed it gently farther and farther away, until its present place was reached.

This distance may still increase, but so slowly as to be imperceptible until after long ages.

" If the moon is the earth's child, it looks very ugly in the earth to push her away!" says some one. So? Even a kind and tender-hearted person will gently push a child away, if it is treading upon them, or crowding them. At first, by its nearness, the moon may have made the earth uncomfortable. Suppose that it constantly pulled the salt tides all over the land! This recession occurs in all satellites, and with the withdrawing, a slowing up of their motion. If we were close to the moon, what should we see? A planet without a drop of water on its surface—not a drop of dew. No place on the earth is so dry as the moon is. The earth has always some moisture in the air, the moon has no moisture anywhere about it. Those flat shining places on the moon surface, once called seas, are really volcanic deserts.

Our earth is wrapped in an atmosphere some one or two hundred miles thick. The moon has no atmosphere. Close to its surface there may be some gaseous layer, but there is not a waft of air, not a fleck of cloud or mist. As there is no dust and no vapor, the moon has no splendidly colored clouds, no blue skies, no varying tints. Clear light is poured from the sun, but there is no diffused light ; no life ;

above it or around it, the moon has only black depths in which shine and flame stars, comets, planets, the sun, the great star. There is no sound to be heard on the moon; no click of moving sand; no jar of falling rock; no rumble of thunder, all is profound stillness. Think of it, a world airless, waterless, soundless, colorless, lifeless. To us the moon looks bright, with a soft, steady, clear radiance. This is because her surface reflects the light of the sun, and this light, tossed toward us, is caught and reflected by the dust in our atmosphere as a beautiful, steadfast illumination.

There is no decay, no downfall, upon the moon, because air and moisture, the chief agents of decay and change, are absent. We discovered that the moon had no water when the telescope showed us that there were never the least clouds about the planet. The evaporation of moisture will always produce cloud or mist. The lack of moisture was easily shown. How did we find out about the lack of atmosphere? When a star is hidden from the earth by the passing of any heavenly body between the star and the earth, that is called an occultation, or hiding of the star. The moon sometimes passes between us and a star. When that happens the star is suddenly lost to sight, and as suddenly reappears on the other side of the moon's disc. If the moon had an

atmosphere, the star would be dimly veiled by that,
before it was quite hidden by passing behind the
moon. How simple some of these explanations seem
after they are made. The nearness of the moon, affording so clear observations of her phases and her
eclipses, early made her an object of interest. What
is that distance or nearness? We call two hundred
and forty thousand miles a long way. So it is, but if
we could traverse it by steam we could reach the
moon in less than a year. If a stone could be
dropped from the moon, and set free of its attraction
of gravitation, it could tumble down to the earth in
three days, or perhaps four days. It would take
longer for it to tumble back, because the gravity of
the earth is greater, and it would not fall so fast at
the beginning of its journey, nor be pulled so quickly
at the end.

The ancient philosopher and teacher, Pythagoras,
who lived five hundred years before Christ, seems to
have had views far in advance of his day, on the
subject of astronomy. It is said that certain sages
of Egypt and Babylon were his teachers, and that
he understood that the sun was the centre of our
solar system, and that all the planets revolved about
it, the moon only moving about the earth, instead of
all the heavenly bodies, as was then currently supposed. Pythagoras taught orally, and he did not

think it well to reveal his views of the world-system to people in general, but merely to a chosen few of his most intelligent pupils; thus only a very few heard from his lips that the sun is the true system-centre. A hundred or so of years later, men who had learned from the favorite disciples of Pythagoras, announced the doctrine of the sun as the centre of the universe. They were at once persecuted for impiety, and their lives were threatened, so that the teachings of Pythagoras were suppressed, and the system later taught by Ptolemy was the only one advocated.

Pythagoras had tried with the rude instruments known in his time to measure the distance between the earth and the sun, and the earth and the moon. He concluded that the sun was nearly thirty thousand miles off, and the moon about fifteen hundred. Proper measurements could not be taken in those early times, and it was not until 1769 that accurate knowledge of distances in our system was obtained. It has taken the observations of many men during many years, to arrive at facts which we speak of lightly as a matter of course. "One man's work for a thousand years could not duplicate them," says one astronomer, yet all this knowledge is now our heritage.

CHAPTER XIII

"The young moon has fed
Her exhausted horn
With sunset's fire."

THERE are many things we ought to know about
our night queen. It is easy to see that the moon is
a round body and travels about the earth, always at
nearly the same distance, and with the earth goes
around the sun, but how shall we clearly explain the
moon's changes?

The new moon first appears to us as a thin crescent, perhaps one-third way up toward the zenith.
Just before sunset we see that silver bow, and perhaps we wonder why, since it is so high, we did
not see it yesterday. Notice that the concave side
of the crescent is always toward the east.

The moon, we know, travels eastward, as the earth
does, and when we see this crescent, which we call
the "new moon," she has just passed the sun, and is
so turned that only a little part of her disc catches
the light of the sun. The changes of the moon
from first, second, and third quarter to full are
called "phases." We shall find that Venus and

128

Mercury, the two planets lying between us and the sun, have also these phases.

The moon continues to advance along her orbit, and when she is ninety degrees from the sun we see one-half of her face illuminated. When twice that distance, or one hundred and eighty degrees, have been gained, we have full moon; she is then in opposition, or exactly opposite the sun, and thus shines splendidly in the east, receiving the sunlight full upon her whole face. We see the full moon rising at sunset in the east.

THE PHASES OF THE MOON.

Still the moon moves on, and now begins to lose what she has gained. She is no longer full face to the sun, but has so moved on, and begins to grow what we call gibbous, a portion of her disc lying in shadow. Each night some of the brightness seems to be cut off, and each night her rising is later, until she shines only in the morning twilight; then passes out of sight in the daylight, drowned in excess of light, suddenly to reappear as a crescent, high up in the west.

Why does not the earth cut off the sunlight

9

from the moon when the moon is opposite the sun, on the further side of the earth, seems here to be a pertinent question. It would do so were it not for that tipping to the plane of the ecliptic of which we have so often spoken. This is the tilting of the orbit in such fashion that the sunlight strikes just above or below the earth, and so can fall fairly upon the moon. It is only planets lying between us and the sun that can exhibit to us these phases.

A question often asked is why the moon turns over so slowly? Sir Robert Ball tells us that the slowness is caused by tidal action, and that tidal action is created in the moon by the attraction of the earth. Once upon a time the moon was not a globe of burnt-out volcanoes, as at present, she was a soft, yielding mass of hot material. In that soft mass the strong attraction of our earth created great tides. That was long, long ages ago, while our world was building into its present condition. The tides excited on the moon were not the gentle, changeful tides we know on earth, because the body creating them was so much greater, and the body in which they were created was so much smaller and in a softer state. They were tremendous tides.

If the moon had rotated too slowly, those terrible tides would have taken her in hand and dragged her furiously along to make her keep up with the earth.

making one rotation while she made one circuit. If the moon had been in a hurry and had begun to tumble over and over in swift rotations, making several turns during her orbit, these tides, being created by the earth, and not mere private affairs of the moon, would have held her back, as by the laying on of strong hands, so that she could make but one rotation while going around the earth. As these tides were raised by the earth they were in their action tied to the earth.

Since the moon has grown cold and hard, and there is not this wild wash of tides upon her, why does she keep up this way of moving?

It has been suggested that although now there are no tides on the moon, there may still be strong tides in the moon. Also there is always a tendency toward a state of permanence, or a resting point, for long ages before changes again become apparent. Nature's laws work slowly and cover periods long beyond our imaginations.

The fact of but one rotation during one circuit causes the moon to present the same face to the earth at all times. This is easy to state, but it is not easy to understand at first.

Many illustrations have been offered, some of which will here be given. This is one : Suppose a boy started on the line of the equator and moved entirely around the earth. We will suppose him to be a boy able to float gently along in the air, superior to such little matters as seas and mountains. This boy is flying a kite. When one flies a kite, only the front side or face is turned to the one who holds the string; he does not see the back of his flying kite. The moving of the boy around the world, as he flies his kite, will cause the kite to make one circuit of the globe also, and during that circuit only the face of the kite will be seen. The moon is the earth's kite.

Here is another illustration: Hold a pencil perpendicularly. Let the point represent the earth. Let a silver dollar represent the moon. Put the dollar against the tip of the pencil, the nose of Liberty pointing to the pencil. Then turn the dollar slowly about, so that it moves around the pencil tip, and still it is the nose of Liberty that is directed to the pencil. If there were a hole through the dollar and a peg through the hole you would see clearly that as the dollar is passed around the pencil it had turned once on its axis.

Again: Take these two apples; let the larger be the earth. Put a hat-pin through it, for an axis to turn it upon, and tilt that axis. The little apple, also on an axis so it can turn over upon itself, represents the moon. Put this small peg in the side of the moon-apple that is toward the earth.

Now move your moon about the earth, letting it rotate on its axis once. When it is one-quarter of the way around, turn it over one-quarter. When it is one-half, turn it one-half. If you tilt your moon-axis properly and imagine the knob on a near chair to be the sun, you can also see how the sun's light will strike the moon in its path, causing its phases. Now, when you have moved your moon around the earth it has rotated once in the circuit, and that peg has been turned to the earth all the time.

By trying such little illustrations one will finally see clearly that if the moon rotates once while it travels its orbit once, it must keep the same face earthward. Now we begin to see it, and this is one of the ideas which, rising in the mind, gain form and clearness by contemplation.

Let us pass to the consideration of whether the moon has ever had any atmosphere. It is supposed that as all other planets seem to have an atmosphere, the moon must once have been so provided. Then how did the moon lose its atmosphere? It is explained thus. Air is chiefly composed of two gases, oxygen and nitrogen. It has small quantities of some other gases. These gases are made up of tiny atoms called molecules. As the great stars are larger than fancy can compass, so molecules are smaller than fancy can conceive. Molecules are very nimble; never quiet; they are always dashing about with great speed. Some of them seem to become excited, and rush beyond the rest, as does the swiftest runner in a crowd. These active little molecules about the moon took frantic excursions, and the moon's mass is so small that her power of attraction cannot pull back bodies which travel off at the rate of a mile a second. When molecules departed so recklessly as that, the moon did not bring her truants home.

By degrees all her molecules went off in a terrible hurry and failed to return. The departure of some made it easier for the others. In the lapse of time the moon was left without any atmosphere at all. The attraction of our earth is so strong that none of the molecules of her atmosphere can get away from her.

This theory is held by very eminent scientists. Some also dispute it, but in so doing offer nothing in its place. As Plato says, " This, then, shall remain our opinion until we find a better."

The moon shines, when full, with so strong and clear a light that it seems as if she must be a light-giving or self-luminous body; but in truth the body of the moon is dark and its splendor comes from that little of the sun's prodigal largess of light which the moon intercepts and reflects. It has been estimated that six hundred thousand full moons would be needed to shine with as great brilliancy as the sun, but the probability is that that is an immense underestimate.

We can prove that the light of the moon is entirely reflected by the very simple method of comparing the moon with the clouds in some early morning, when the moon is still half-way above the horizon and the sun is rising; the moon and the clouds are both reflecting the sun's light. We see on such an occasion that the light of the moon, or her brightness, and the brightness of the clouds, is of the same kind.

The apparent size of the moon is nearly constant. One well-known astronomer illustrates in this wise : " If we place a globe one foot in diameter one hundred and ten feet from our eyes, it will hide the moon. Only very occasionally the globe would need to be brought closer or removed farther off to hide the moon, whether she appeared as full, half, or crescent. If the moon were to drift from us into space, her apparent size would dwindle; if she came rolling nearer, her apparent size would increase."

Although the size of the moon is constant, it is very remarkable that she seldom appears of exactly the same size to any three persons observing her at the same time. " How large does the moon appear to you?" we ask, and a startling variety of replies comes: "As large as a coffee saucer," "As large as a dinner plate," "As large as a car wheel," "Oh, no, as large as a cart wheel," and so on.

There is no time on record when the moon was nearer the earth than now; her present position and her present consequent apparent size were attained long before earth furnished any moon-gazers. It is evident that this constant distance of the moon from the earth could only be kept if the moon revolved about the earth, for if the moon did not so journey around the earth, the attraction between the two bodies would bring them together. As the earth has been set traveling about her greater neighbor the whole power of the attraction is expended in keeping the moon in her nearly circular path. If this attraction were suddenly released, the moon, instead of journeying in her orbit about the earth, would begin to drift back into space and never return; in a comparatively short time the moon would be forever out of sight, and we should have no more bright-light nights. Then what would poets, painters, and lovers do?

The moon as well as the sun has its eclipses. The earth shadow creeps over the face of the moon, cutting off the light of the sun, and then the planet becomes dark. There is one remarkably beautiful

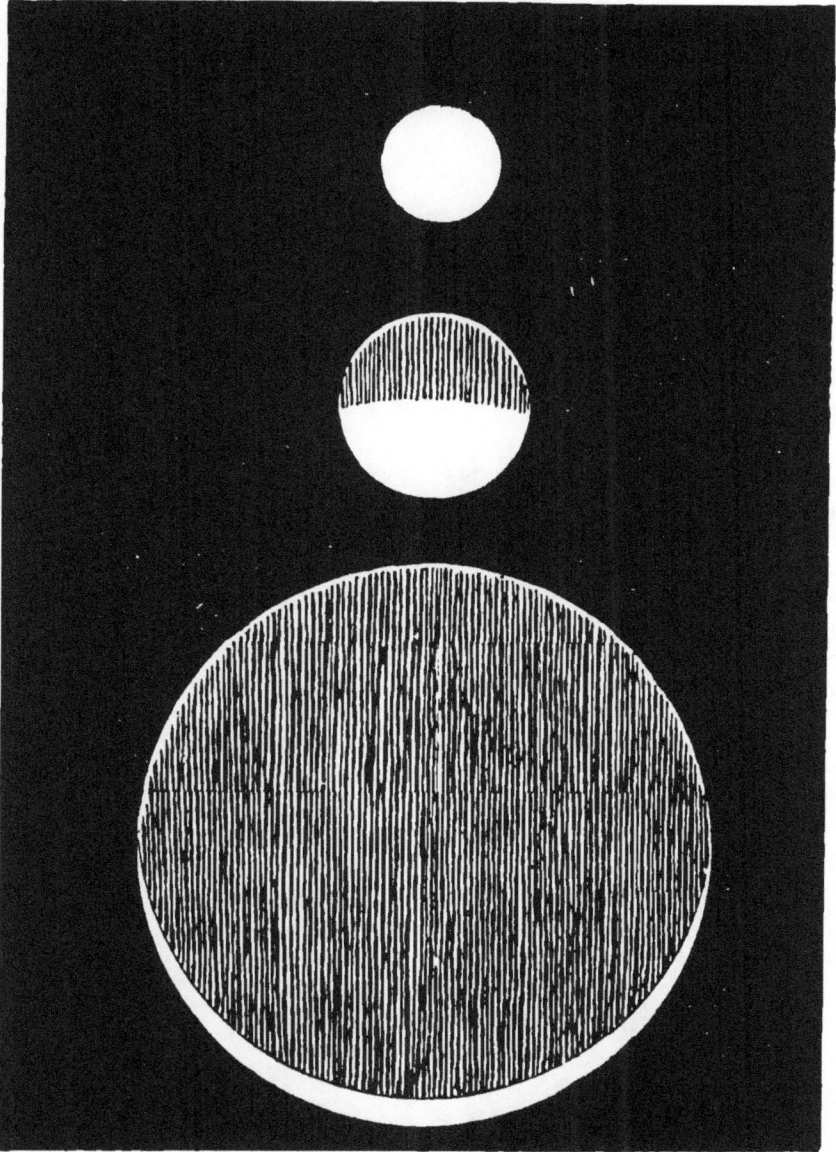

and interesting phenomenon to be noticed in some lunar eclipses. The moon may be so hidden behind the earth that not one ray of direct light can fall upon it, and yet the whole moon-ball glows with a dull tinge, like molten copper. Why is it that we can even in this glow see some of the markings of the moon's surface? This light comes from some of the sunbeams which, having just grazed the edge of the earth, have become bent by the refraction of the atmosphere, and their deflected light thus passing through the immense thickness of the earth's atmosphere, loses its clearness, and takes on this coppery tinge. This effect of our atmosphere on sunbeams can also be seen at early morning, or in the "afterglow" of evening, when the light is more ruddy than at noonday.

Other matters of interest in regard to the moon will be treated under the subject of eclipses.

CHAPTER XIV

WHITE VENUS

"Hanging in a golden chain,
A pendant world."

DAY is closing, and we have wandered over the fields and stopped at the top of a hill to watch the sunset. Above a belt of distant woodlands lies a band of crimson sky, against which rise, white and sharp, the spires of two churches.

Over the vivid flush of crimson are broad stripes of purple, blue, and pale green; just over the deep glow of the descended sun, in a field of delicate saffron tint, shines a great white star. "Oh! what is that splendid world which was not there last night? How has it floated so suddenly out of space?"

It is Venus, the evening star. One of the charms of this star is its sudden appearance, near the setting sun, after having been for some time invisible. With its clear white beams, standing out against the brilliant colors of sunset, it seems like a great translucent bubble. You will see it, near its present position, for some days in great beauty. Venus will be in the west for some time, rising slowly higher, and in two

138

or three weeks it will be high in the heavens in full magnificence, until late at night. Then its light will decline, and the planet will pass out of sight.

It will not be lost, but moved eastward. Some morning, if you will go out on the balcony in the pearly gray light of coming dawn, your great white star, like a lamp hanging low, will be near the eastern horizon, still the most beautiful object in the skies; become the morning instead of the evening star. You will lose it after a time, just as you lost it at first, and it will return as now—the evening star. Thus it will move and change, cycle after cycle.

Venus is the planet next within our orbit about the sun. It is our nearest neighbor, the moon excepted. At some parts of its orbit it is but twenty millions of miles away from us, at other parts it is one hundred and fifty-four millions of miles off. This variation in distance accounts for its apparent changes in size and brightness. Both these distances are enormous, beyond our realization; yet at twenty millions of miles Venus is near enough to often seem the most splendid of the hosts of night.

Venus is the most nearly round of any body in our solar system, and has the most nearly circular orbit. On account of its being so much nearer the sun, the sun seen from the surface of Venus would appear twice as large as to us.

On Venus night and day are practically of the same length as ours, for the planet appears to rotate in about the same time as the earth. Its orbit, being shorter than ours, is traversed in two hundred and twenty-four days.

Venus is·more tipped, or inclined to the plane of its orbit, than our earth. As it is this tip which regulates the seasons, the seasons upon Venus are more sharply marked than ours. The winters are of about the same coldness, but the summers are very much hotter. Venus has no temperate zones as we have on the earth; the frigid zone extends to the torrid, but is not so fiercely cold as our polar circles. There is no polar ice-cap on Venus; the shortness of the winters and the intense heat of the summers do not permit an accumulation of snow.

Some astronomers think that mountains have been seen on Venus, measured and found to be twenty miles high. Other astronomers are far from sure that this is yet proven. We understand the seeing and measuring of moon mountains, the moon being nearer seems so much larger, but how could they see and measure the mountains, if there were any such, upon a planet which distance makes so small? Here we come upon some wonderful facts about Venus. Venus has phases or changes just as the moon has. The planet appears to us

sometimes full, sometimes as half-round, sometimes as a crescent. It is upon the inner edge of
the crescent that mountains have been supposed to
be seen. Of course, the phases are never noted by
the naked eye, for the image of the planet cast upon
our eyes is very small, too small for any such
changes to be visible. The telescope shows us these
phases.

The orbit of Venus is within ours, and as it journeys about the sun, within the path that our earth
describes, it comes between us and the sun, and
then its dark part is toward us, and its illuminated
face is toward the sun. When it is to the right or
left of the sun it presents, according to its position,
but a half or a quarter of illuminated surface.
When at last it moves around to the other side of
the sun it shows us its whole shining disc. Galileo
first observed these phases in September, 1610. Such
are rare, glad moments in an astronomer's life.

If Venus is between us and the sun, why do we
not have an eclipse of the sun from Venus as we do
from the moon? However, in these affairs of the skies
distance is as important a factor as nearness. We
have no eclipses of Venus. Venus is so distant and
so small that it could never have sufficiently great
apparent size to obscure the sun's disc. Owing to the
nearness of the moon to us its apparent disc is as

great as that of the sun. The passing of Venus between us and the sun, across the sun's face apparently, cannot occasion an eclipse or hiding. It is called instead a transit or crossing. The small, dark body of the planet travels across the sun's disc, and in this progress has been eagerly studied through telescopes. This is a most interesting spectacle, but we may not propose it for immediate observation. We would need to watch a long while. There will be no transit of Venus until 2004. The last one was in 1882.

If the planet is moving along between us and the sun why is there not a transit every year?

The time of the planets is different. Venus and the earth do not move side by side, like a pair of horses in harness; rather like two racers on a course, one with a larger circuit, or greater swiftness, than the other. Venus makes thirteen journeys while the earth makes eight. Making these journeys the earth overtakes Venus once in nineteen months. In that case why is it we may not have a transit that often?

Owing to the inclination of Venus to the plane of the earth, in half its journey Venus is above our orbit and in half below it, and thus it usually happens that when our earth "catches up" with her brisk neighbor, Venus is above or below the face of the sun, as it appears to us. When one transit has

been made, in eight years there will surely be
another, and then not one for perhaps over a hundred years.

We must not think of planet tracks as ring within
ring, but as of loops of cord, lying not exactly true
to each other, as if we drop the cord in this wise:

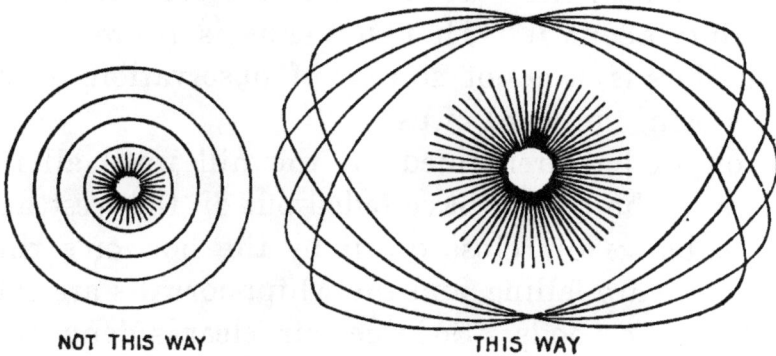

NOT THIS WAY   THIS WAY

How large is Venus?

Its diameter is only two hundred and fifty miles
less than that of the earth. Venus is less heavy
than the earth, being three-quarters of the earth's
weight.

Venus in not only a beautiful object in the skies,
but its transits have afforded us a scale of measurements for the other bodies of our system, also one
very good way of determining the distance of the
sun from the earth.

Venus has an atmosphere; we do not know how

deep that atmosphere is, or of exactly what gases it is composed. There is evidently moisture upon the planet, in lakes, rivers, or seas, for there are clouds in its atmosphere, often beautifully colored, as in our sky.

So far as is known Venus has no moon. Astronomers have thought they discovered one satellite lingering near the beautiful star, but there is yet no certainty about it. Though Venus is nearer to us than Mars is, it is not so easy of observation, owing to its being sunward of us.

Now we have remained on the hill until all the colors of the sunset have faded out of the west, and Venus hangs low, close down by the horizon's rim. The dews are falling, and the whippoorwills are calling from the woods. So large, fair, clear is Venus that we might fancy our earth had in her a second and very small moon.

As we watch these heavenly bodies it is interesting to consider how life upon them must be attended by conditions very different from anything which we know in our world.

There is little likelihood that man could exist on any other orb than his own in our system; the earth has been prepared for him and man for the earth. The atmosphere surrounding our world secures for us a suitable temperature; the period of rotation upon its axis secures for the earth the change from day to night which man needs for rest and quiet; the period of the earth in its orbit gives us the seasonal changes. In our system there are globes of fluid or gaseous fire as the sun, Uranus, Neptune; there are frozen globes such as our moon, with its day a fortnight long; there are globes of tremendous conflict and storm, such as Jupiter, which also has a day and night of five hours each, and a year twelve times as long as ours.

Some astronomers consider that Mars is the one planet where life conditions are so similar to those of our world that, while people constituted exactly like human beings could not survive if transplanted there, it is yet easy to conceive of beings with somewhat modified human conditions having there a home. Mars has day, night, seasons, atmosphere, and probably water; some types of animal life might surely be so constituted as to thrive there.

But how about life on Venus? Why should not the beautiful planet be a fit home for beings so nearly like ourselves that we might claim them as kin? The size of Venus is nearly that of the earth, being but two hundred and fifty miles less in diameter; its density is three-fourths that of the earth, and therefore the attraction of gravity is one-fourth less. We illustrate by saying that any body that here weighed one hundred pounds would there weigh

10

but seventy-five; a force that would enable one to jump forty inches up here would carry one fifty inches on Venus; a body let fall here moves sixteen feet in a second, on Venus it would move about thirteen.

While the time of the rotation of Venus is not absolutely fixed, it is supposed to be about forty minutes less than that of the earth; thus the happy interchange of day and night is assured there as here. The climate of Venus would be very much hotter than we experience in our world; yet, as in the small circuit of our globe we find the luxuriant vegetation of the tropics born in seething, moist heat, and the small lichen vegetation of the far North, thriving on and under beds of snow; when we find man close to the poles, journeying in sledges, and dressed in furs, and running nearly naked in equatorial forests, we see that animal and vegetable life can support and adapt itself to singular extremes of heat and cold. We then consider that plants and animals may be fitted for the hotter year of Venus, which is, at least, less violent than ours in its seasonal changes. Water and an atmosphere Venus certainly has; if oxygen is present in that atmosphere there may be on the planet life largely like our own.

Leaving speculations and returning to facts about Venus, we find that Venus shares one peculiarity of the moon in being visible sometimes to the naked eye in daylight. As has been said about the moon, the white splendor of the planet is so great that it often seems as if it must be giving out an effulgence of its own, and yet really it is a dark body, such as the earth or the moon, splendid only in the radiance it gathers from the sun and reflects from its inert surface. We may take the method before used with the light of the moon, comparing the light of Venus in daytime with the kind of light shown by brilliantly lit clouds, and we see it is all reflected, not self-given.

The nearness of Venus causes it to be from forty to sixty times brighter than any orb in the northern sky, and so bright that it can be seen sometimes, as the moon is seen, shining in high heaven, and boldly challenging comparison with the light of day.

One might long and pleasantly continue discussion of the questions and discoveries which cluster about this beautiful planet, which has had so much time and labor devoted to its phenomena. The planet has been to us a fruitful source of information, it has been as a golden measuring-rod whereby we have tested the distances and learned the unspeakable majesty of the solar system.

CHAPTER XV

"Yet 'neath a curtain of translucent dew,
    Bathed in the rays of the great setting flame,
    Hesperus, with the hosts of heaven came."

WHY are we not all content to live our own lives down here and let questions about the solar system rest?

In all these studies in Natural Science it is that we "think God's thoughts after Him." That ennobles and enlarges the mind, and to these thoughts and studies the normal mind is usually drawn.

What thought are we going to think after our little study of the white star Venus?

What we might call a hidden thought, Mercury, so far as we know, the youngest of the solar family, the little child planet, set close by its father's side. Long ago the ancient star-gazers discovered, near to the setting sun, seen only when the horizon was clear of clouds or mist, a small pale-pink planet, which they named Mercury, after the god of thieves, saying that this star rose to show the thieves when to begin their work! Those old astronomers also

148

saw close to the rising sun, when the morning was fair, a small pale-pink star with a clear beam. This star they named Apollo, after the god of day.

Really these two stars were but one; the planet nearest the sun appearing sometimes as a morning and sometimes as an evening star. It was only after many centuries that it was fully ascertained that what had been called two stars was merely one, and for that one the name Mercury was retained. Ever since that time this lonely little planet oscillating about the sun has been an object of curious interest.

Mercury shines very brilliantly sometimes when everything is favorable to its appearance. It is about one-fourth as large in diameter as our globe, but less than one-twentieth of the earth's weight, and is a little over one-third as far from the sun as we are. Our orbit is some ninety-two millions of miles and the orbit of Mercury thirty-five millions of miles away from the fountain of light and heat.

This little planet rotates upon its axis once in twenty-four and one-quarter hours, so its day is about the length of one on earth. It completes its short orbit in eighty-eight days.

The mass of Mercury is not quite one-fifth that of the earth, but the planet is three times as dense or solid as ours is. Upon this compact little globe the sun pours nearly seven times the light and heat

that is afforded to us. A French fable says that all
the inhabitants of Mercury are mad—that is, crazy—
because of sunstroke! We might fancy that the
planet itself was mad from overheat, it shows so
much eccentricity in its orbit. As Mercury shows
the most eccentricity in orbit and Venus the least,
we may look on Mercury as the spoiled youngling
of the family of the sun and Venus as the model
child. Eccentricity in orbit means variation from
the circle. By so much as a planet's track varies
from a true circle by so much it is eccentric. Mercury is the most eccentric of the planets, unless we
except one or two asteroids, which are supposed to
be very eccentric in their travels.

If a planet revolved about the sun in a true circle,
then at all periods of its year it would receive equal
light and heat. The circle being flattened to an
ellipse, by just so much as it is in one part flattened
or pressed toward the sun, the other parts are extended or drawn from the sun. As this is a matter
of perhaps millions of miles, in what is so great as a
planet's orbit, a perceptible difference is made in
the amount of heat received.

Mercury, revolving about the sun, reveals to us
phases as Venus does. Sometimes Mercury is dark
to us, the sun illuminating the side that turns from
us sunward. As it moves to right or left of the sun

we get a half, or quarter, or eighth of the planet illuminated, and finally it smiles to us, in full light, on the other side of the sun. Of course these phases can be seen only through a telescope. Admirably delicate as is the human eye it is yet too coarse an instrument to detect such fine points as these. The object cast on the retina is too small for it to grasp. As is the case with Venus, some observers think that they have discovered very high mountains jutting from the crescent Mercury, but this is disputed by others.

When can we see Mercury best, as it lies so near the sun? If the weather is clear at morning and evening we can see Mercury in March and April, and again in August and September, Look for the star in the early twilight, morning and evening. Get your planetarium, that will point it out exactly. So far as we know Mercury has no satellite. The sun is his nearest companion and in the glory of the sun he is usually lost to mortal eyes.

Some think that Mercury has an atmosphere, much more dense than ours, but nearness to the sun makes the planet so difficult to observe that much about it is conjectured rather than proven.

Have we ever stopped to think that the speed of every planet differs in different points of its orbit? At the points in its orbit nearest the sun the planet whirls forward swiftly; farther away it slackens its pace somewhat. The planet, like a horse, weary in the length of its journey, falls into a walk, having gone far from home, but picks up spirit and moves briskly again when home is once more in view. The reason of change in velocity in the movement of the planet is the attraction of the sun; of course, greatest when the planet sweeps nearest.

We should find living hard work on a planet six times hotter than our earth, we think, when wilting in summer heat. Mercury must be at least nine times hotter than our earth. Its density may cause it to receive, store up, and radiate much more heat than our globe. Take, for instance, a pound of cotton and a pound of iron and heat them before the fire. The cotton—unless it takes fire and burns up —cannot absorb much heat on account of its loose texture, and will never feel more than warm to your hand. The iron will absorb a deal of heat, because its texture is so compact. It will not only burn your hand if you touch it, but will make it very warm if you merely hold it near.

Similarly, the surface of the dense planet Mercury receives and radiates much heat. The climate of a globe does not depend entirely upon sun-heat; atmosphere modifies it. The atmosphere about a globe may repel or retain much of the sun's heat.

Knowing so little of the atmosphere of Mercury, we cannot be sure of its climate.

Mercury makes its transit across the sun, affording opportunity of careful study. During this transit a shadowy circle about the dark body of the planet seems to suggest a dense atmosphere.

It has been thought that the eccentricity of Mercury in its orbit might be due to some other planet, not yet discovered, lying between Mercury and the sun. Such a planet has been searched for under the name of Vulcan, the god of fires. This interior planet has not yet been seen. If it really does exist it cannot fail at some time to come in line between the earth and the sun, and then will be seen making its transit across the glowing face of the sun. That will prove that Mercury has a smaller and younger brother. Some astronomers think that Venus and Mercury rotate only once in their trip around the sun. That, however, is not proven.

There are some countries in the world where the planet Mercury is seldom or never seen, and this is not owing to their position, but to the dampness of their atmosphere causing hazy or cloudy sunsets and sunrises. The famous astronomer Copernicus knew of the planet Mercury, and greatly longed to see it, but died without getting a view of the very planet which was the most striking example of those celestial motions which Copernicus first clearly expounded. The vapors in the atmosphere about Frauenburg, the home of Copernicus on the Vistula, always prevented the eager astronomer from seeing the antic little planet. An astronomer so situated in these days would promptly step aboard an express train or a steamship and visit an observatory in some more favorable locality. Traveling was slow, dangerous, difficult, and costly in the days of Copernicus, and he had to give up viewing Mercury, until, as Galileo said, " he passed it in his flight among the stars after leaving this world, and then knew what was true."

Mercury was doubtless a less easy matter of discovery than other planets known to the ancient world. It seemed, as we have said, not one but two planets, and all conditions for observation must be exceedingly favorable, or it could not be seen at all. Again, none of the planets is so variable in appearance as Mercury when at last it does happen to be visible. Both its height above the horizon and its brilliancy vary greatly, so that it seems not possible that such different appearances should belong to one planet. How many ages, and how many thousands of observations must have been required to establish the fact that these double appositions, these revealings of a morning and evening star, of a brill-

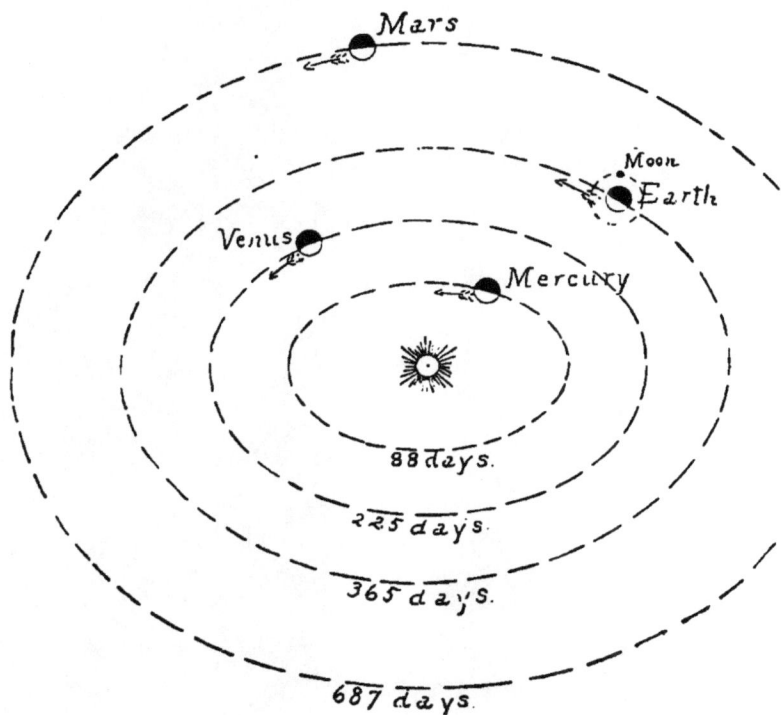

Mars

Moon
Earth

Venus

Mercury

88 days.

225 days.

365 days.

687 days.

THE FOUR YOUNGER CHILDREN

iant and a pale orb were one and the same; that one of these close dwellers by the sun was never seen until the other had gone, that both were never in the field of vision at the same time, while they occupied nearly the same place. Learning that Mercury was not two, but one, seems a brilliant astronomical discovery.

In the British Islands, Mercury can usually be seen during the course of the year; in the United States, owing to the dryer, clearer atmosphere he is freely observed, but in the clear skies of the desert of Arabia or Africa, must have lived the star-gazers who were able to trace the wanderings of this sun-child and name him long ago.

Mercury, being such an intractable little planet with which to make acquaintance, we were obliged to apply to a comet to learn his mass. How was that? It was early known that Mercury was much less in size than our earth, and was also much more dense. Of late a new estimate both of the size and density of the planet has been reached, which the astronomer C. A Young considers more likely to be trustworthy than those previously fixed upon and given in most books. In the year 1895, when Encke's comet was making its flight in our system, between us and the sun, it passed, as is its wont, very near to Mercury. No other sky wanderer, no heavenly orb, ever gets so near to Mercury as Encke's comet. Now Encke's comet is less active than other comets, and is almost structureless, it seems of very tenuous material, "like a smoke wreath." Upon this comet the compact little planet Mercury exercises much influence, and by the perturbations which it creates in the comet, the learned can estimate the mass of Mercury. As a consequence of observations and calculations made in 1895, Backlund, the astronomer, and others, conclude that the mass of Mercury is only one-thirtieth that of the earth, and its density about two-thirds. This shows us that Mercury is very much more solid and compact than our earth, and if it were as large as our earth its attraction of gravitation would be immense, probably twenty times as great.

If the usual question of fitness for habitation comes up concerning Mercury, we can only say that no beings of which we could form any conception could abide there. Dwellers on Mercury would need to be those whose native element was something very like fire, for upon Mercury the solar heat streams forth as upon no other planet of our system.

CHAPTER XVI

ECLIPSES AND WHAT THEY TELL

"Or from behind the moon,
In dim eclipse, disastrous twilight sheds
On half the nations, and with fear of change,
Perplexes monarchs."

THE "Columbian year" was rife with tales about the great discoverer, and one which especially charmed many children ran about an eclipse, in this wise : "In 1502, when Columbus was at San Domingo, the natives would not allow his men to land to get water, and would not furnish him with provisions. He knew that he must have supplies, and he frightened the natives by telling them that the Sky Spirit was angry with them because of their inhospitality, and would take away the moon that night. This as he knew that there was to be an eclipse. When the people saw the moon disappear they were terrified, and gave Columbus whatever he wanted." Fiction abounds in the use of similar incidents as part of the machinery of tales. All ignorant and savage people have feared eclipses. They supposed them to be signs of the wrath of the gods, and to foretell

157

great disasters. As the theory of eclipses became plain, it was found that they could be calculated not merely a month, or a few years in advance, but many years—even centuries. If you wish to know when there will be an eclipse, we say, as Bottom said in Shakespeare's "Midsummer Night's Dream": "A calendar! a calendar! Look at the almanac. Find out moonshine! find out moonshine!"

However often eclipses occur, they never fall into the category of the commonplace. There are solar and lunar eclipses. The highest number possible in one year is seven; the lowest is two. When there are seven, five will be sun eclipses, and two of the moon. If there are but two, both will be solar. The usual number in a year is four. We have noted in a previous chapter that an eclipse of the sun was caused by the coming of the moon between the earth and the sun, so shutting out the light of the sun from us. An eclipse of the moon is caused by the moon being in opposition—that is, on the side of the earth opposite the sun, and so receiving its full rays, and our earth cutting off that light by coming between the sun and moon, and casting her shadow upon the moon.

An eclipse may be total, or partial. An eclipse of the sun must always occur when the moon is "new;" an eclipse of the moon can only take place when the moon is " full." The reason for this is, that the moon never comes between the earth and the sun to afford an eclipse except at the period called " new moon ;" and she is never in opposition to be eclipsed herself except when at the period of " full moon." In the total or partial shutting out of light, called an eclipse, the sun is darkened by the moon, the moon by the earth.

To make the theory of eclipses plainer, let us talk a little about shadows. When light is thrown upon any opaque body, that body casts a shadow. Every planet, big or little, casts a shadow in the direction opposite to the sun. The form and size of these shadows depend upon the comparative size of the sun and the planet, and their relative distance. For a practical example, let us take some oranges, and let us illustrate the law of shadows. Here are two oranges of equal size. Let one represent the sun shining upon the other one; half of the one is lit up, its other half lies in shadow, and that shadow will fall in the shape of a cylinder. If there were a planet equal in size to the sun—which there is not —the shadow cast into space by that planet on its night side would be cylindrical.

Take away the orange that we called a planet, and choose a small one. The shadow cast by the small body, illuminated upon one side by the sun, is not a cylinder, but a cone. The shadow tapers off until it ceases altogether. If the planet were greater than the sun, the shadow would not be cylinder-form, it would not diminish as a cone, it would expand. The nearer the sun to the great planet it illuminated the more widely the lines of shadow would diverge.

As all the planets are smaller than the sun, each casts a cone-shaped shadow.

If the moon's orbit, or path, lay exactly in the plane of the ecliptic, the moon would eclipse the sun every month. Just here we might explain what is meant by the frequently used phrase, " the plane of the ecliptic ?"

The ecliptic is a vast imaginary circle drawn in space, having the sun for its centre. Every part of this circle lies true, or level, to the centre of the sun. Imagine this great sun-centred circle ; now fancy that you can lay a long ruler upon the lower edge of the ecliptic, or circle, and push it steadily forward until it passes through the sun's centre, and then on, and finally off the upper edge of the circle, the ruler lying true to the circle all the time, and in its progress touching every part of it. That is the idea of the plane of the ecliptic ; or, fancy a sun floating half buried in water, and let the water surface be the plane.

If the orbit of the moon were true to this plane the moon would pass before the sun each month, and cast her shadow upon the earth, so that we should lie in her shadow and see her dark form against the sun, shutting out our light. The moon's orbit is, however, tilted about five degrees from the plane of the ecliptic, and therefore at new moon she may be between us and the sun, or she may be above or below the line. Also at full moon she may be above or below the line again, and so not be eclipsed herself by the shadow of our earth. Thus:

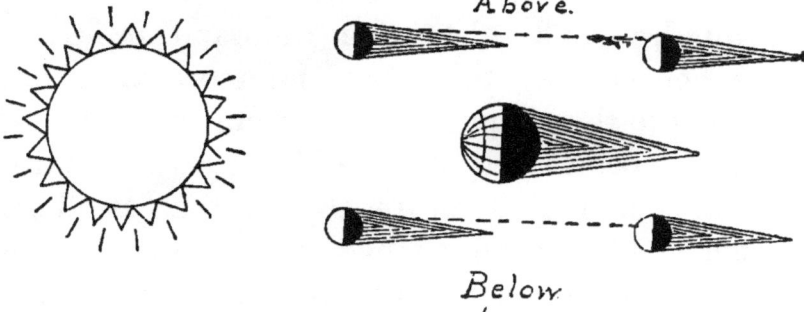

An eclipse of the sun comes on from the westward, the shadow moving across from the sun eastward. In an eclipse of the moon the shadow falls upon the moon on her eastern side, and moves over

11

to the west.   This is because of the eastward motion
of the moon in her orbit.   .

How long is the cone of shadow which our earth
casts?

About eight hundred and sixty thousand miles;
more than three times the distance from the earth
to the moon.   The average breadth is some six
thousand miles.   It is wider close by the earth, and
narrower farther along.   When it reaches the moon,
its breadth is about six thousand miles—that is,
three times the diameter of the moon.   Plenty of
room to lose the moon in that shadow!

The length of the moon's cone of shadow averages
two hundred and thirty-nine thousand miles; it
varies between less than that and more than that,
according to the moon's distance from the sun.   As
the moon's orbit is elliptical, and the earth's is also,
she is at some periods nearer the sun than at others,
and her shadow cone changes its length with the
distance.   Sometimes it is two hundred and fifty-two
thousand six hundred and thirty-eight miles, or
more than twelve thousand miles longer than the
distance between the earth and the moon.   Its
breadth at the distance of the earth is only one hundred and seventy miles.   The lighter extent of shadow
beyond the very dark portion is, however, sometimes
over four thousand miles in width.

The dark shadow of an opaque body is called the *umbra*, a Latin word, meaning a shadow. All portions of a shadow are not alike dense; on each outer edge of the shadow there is a portion of lighter shadow, called a *penumbra*.

A partial eclipse of sun or moon is when the intervening body passes over but part of the eclipsed disc. When the earth's shadow falls centrally upon the moon, the moon is always totally eclipsed. An eclipse of the sun is central when the moon's shadow falls full on the sun's centre; it will be total if the moon is so placed in regard to the earth that her disc is apparently as large as that of the sun. If her apparent size is less than that of the sun, the eclipse will be annular, a ring of the sun's glowing disc showing all around the moon's shadow.

While an eclipse of the moon can of course be watched by the naked eye, and the general public frequently views a solar eclipse with great satisfaction through a sheet of smoked glass, it is in observatories that the eclipse awakens an intense interest, and in them all are at work to learn what they may from this " revealing by hiding."

For some years past there has been a revival of interest in the subject of astronomy, and an unusual activity in the building of observatories. As we saw in the case of Camille Flammarion and his observatory, the wealthy amateur often comes to the aid of the learned astronomer, providing him with the requisite tower or instruments, while colleges are now demanding a far better apparatus than once would have contented them.

Telescopes are now constructed with lenses of such immense size as once would have been esteemed a wild dream. Nations or individuals are choosing sites which would be particularly favorable for star study, and there providing an astronomical equipment adequate to the desired end. Thus Arequipa in South America has an observatory which enjoys a peculiarly clear atmosphere on its Andean elevation, and Arizona, in our own country, has an observatory as a station auxiliary to Cambridge, as in Arizona the dryness of the air permits of work which may be hindered in other places. Greenwich and Mendon in Europe have now telescopes of the largest size, and the Mendon observatory has an adjunct on the frowning and silent slopes of Mont Blanc, where, on special occasions, astronomers may resort for special work. American astronomers, aided by men of wealth, are working in the Arequipa observatory in Peru, and the principal telescope there has a lens with a diameter of eighteen inches. We are told that the work in Peru will be of a kind that cannot be done in Cambridge, because there the sky is all

A TOTAL ECLIPSE

night long illuminated by electric lights. From this
we see that the luxuries of an advanced civilization
have their peculiar disadvantages. The " new astronomy," as it is called, requires for its observations
a pure atmosphere and a sky free of the glare of
reflected city lights, or the outpouring of city smoke.

Italy has always been a land of astronomers, we
might say of astronomers working under difficulties,
for during long ages neither the Church nor the State
aided or abetted astronomical studies. Galileo's name,
of course stands first among the Italian star lovers ;
Cassini, Piazzi, and others range near him.   Pisa furnished the world with a telescope ; Schiaperelli, Padre
Denza, Padre Sais, and others are now doing a grand
work and have been accorded grand opportunities.
The observatory of the Vatican is associated in all
minds with the revision, under Gregory XIII, of the
calendar.   The tower of this structure is called the
Gregorian Tower, a massive domed building having
twenty rooms, where all manner of astronomical work
is carried on.   After the calendar was revised, and the
enthusiastic worker, Ignazio Dante, and his assistants
were dead, the Vatican observatory fell into neglect,
the instruments were not renewed, many were in-
jured or lost, dust gathered on the floors and
windows, and spider webs hung thickly upon the
walls. Now, while a line of spider's web is invalu-

able in some observations, draperies of spiders' webs in observatories suggest that enthusiasm in astronomy has sunk to its lowest ebb. In 1888 the Vatican observatory was remodeled and refitted, given a good endowment, learned and zealous men were called to it, an ancient tower, the Leonine, was added to the observatory and thoroughly equipped. This tower is said to be almost as solid as the Pyramid of Cheops, and was originally built as a fortress against Saracenic invasions. Owing to its massive solidity and position on the Vatican hill, this tower is nearly free from vibrations which interfere with so many observations. Here a very great work has been done in photographing the heavens ; cloud photographs have also been taken in large numbers, and various series of valuable photographs of eclipses—both lunar and solar—have been made.

CHAPTER XVII

THE STORY OF THE TIDES

"The tide rises : the tide falls :
The twilight darkens, the curlew calls ;
The little waves with their soft white hands
Efface the footprints in the sands :
The tide rises : the tide falls."

REFERENCE has frequently been made in these pages to an instrument called the spectroscope. Chemistry brought this instrument to the aid of astronomy. Chemistry is the study of matter, as composed of atoms ; and the study also of the atoms. The chemist analyzes, or divides up, substances, to discover of what they are composed. When a substance can no longer be divided, it is called simple.

Air was analyzed and found to be composed chiefly of oxygen and nitrogen, and these gases were called simple substances. In the course of research and improvements in chemical work it has been found that substances once supposed to be simple can be still further divided into two or more compounds. It is hard to say what is really simple ;

what we call simple this year may be proved capable
of division next year.

Chemistry gave us the spectroscope, and chemistry has applied this wonderful instrument to sun
study. What is called the solar spectrum is an
image of the sun cast in lines arranged according to
what is called wave-length. This divides light into
the various colors of which it is composed, and the
colors tell us in various darkened lines what substance has cast them.

It was found that some substances seemed to be
in the sun which had not elsewhere been discovered.
One was named helium, or "sun-stuff."

Astronomers said if the theory that the various
planets were formed from the sun were true, then substances found in the sun should more or less pervade
the planets. They set about a diligent search, and
found helium in meteoric stones, and also in this
world. This gave fresh confirmation to the nebular
theory, the theory which asserts that all our solar
system was once embraced in the sun.

The spectroscope has assured us of the gaseous
nature of the sun, and also of its capacity for continued heat-giving. Sir Robert Ball gives us a very
fine illustration of this. He says that the sun is a
vast ball of enormously heated gas, and contraction
is as a hand constantly squeezing that ball. As a

sponge full of water pours forth water when squeezed, so contraction squeezes heat from the sun. This contraction, or squeezing, constantly lessens the diameter of the sun. The lessening is at the rate of ten inches each day. So great is the size of the sun that this shrinkage can go on for thousands of years, without causing any noticeable difference in size or heat, except when measurement is applied.

This consideration of the gaseous nature and fierce activity of the sun suggests to us the theory of tides in the sun and tides in all the planetary bodies cast off from his substance. Let us study for a little, tides as known in this world.

The tides are the daily pulse-beats of the sea. Every one who visits the coast has noticed that twice each day of twenty-four hours the waters rise high, and creep up on the beach. Twice each day the sea shrinks and withdraws, and leaves the sands bare, far away. We understand that these tides are caused by the moon. This has been known for hundreds of years.

Long ago there must have been the first man, or first few men, who thought out for themselves this connection between the tides and the moon. We do not know who these men were, but we are sure that they were deep, earnest thinkers, and careful observers to discover all this, unaided by instruments or any knowledge of the great law of the attraction of gravitation; all that we can say about them with certainty is, they must have been people who lived by the sea, and saw its changes every day; they were those to whom the tides meant something personal; they were surely sailors, fishers, or merchants who traveled upon ships to do their trading.

Not only are there two daily tides, flowing and ebbing in regular periods, but the high tides have a period of being very high, and the low tides of being very low. The very high rise is called a spring tide; the very low ebb is named a neap tide.

The attraction of the moon is the chief cause of the tides upon our globe. The sun has also tidal influence, but the sun is so far from us that his attraction is less noticeable. When the attraction of sun and moon combine, that causes the spring tide.

Tides out in mid-ocean are less marked than on continental shores. The island of St. Helena has very low tides. Bodies of water nearly inclosed, as the Mediterranean and some other seas, have very low tides, almost none in fact.

Tides set far back in rivers and inlets; they are noted perhaps a hundred miles inland. At some places the tides rise very high; they press up the Thames to a depth of eighteen or nineteen feet at London. In the British Channel the tidal waters crowd up, swelling against the cliff, so that spring tides rise thirty-eight feet, and neap tides are but ten feet lower.

The highest of all tides are noted in the Bay of Fundy, where the rise and fall of a spring tide is fifty feet. This mass of water, gathering in from the open sea, comes with a great sudden swell or bore. There is a distant, deep, warning sound, as the roll of thunder in a heavy storm. The hogs, which go down to the beach to feed on mussels, oysters, and crabs, lift their snouts from rooting in the mud, and gaze earnestly seaward. Then they jerk up their heads high with a " woof! woof! woof!" and away they go, pell-mell, as fast as they can race inland, until they are well out of the way of the advancing water.

The rise and fall of the tides causes strong currents; these can be used, as any other water power, to drive machinery.

Tides have cut down banks and carried away large tracts of country. On the other hand, tides have brought material and formed banks and new levels. Tides have had much to do with carving the outlines of the land, cutting out inlets, bays and straits. Far out at sea the tides cause currents which are very useful to navigators.

There is one very interesting fact which we may not at first blush be able to understand—this is, that the tides draw upon and lessen the stock of energy in our earth. As this is lessened, the earth rotates more slowly upon its axis, and thus the day is lengthened.* This process is so slow that many thousands of years must pass before any visible difference would be made.

Tidal influence reacts upon the moon, and has driven the moon farther from the earth since her formation. No doubt this retreating of the moon may still go on to some slight degree, but the present distance will not be appreciably changed for many ages.

The creating of tides upon one planet by the attraction of some other planet extends through all the solar system. Many of the planets have not yet reached a stage where there is any water upon them, but water is not necessary to tidal motion. Tides can be created in a body that is all of heated gas, or is in a fluid or semi-fluid condition of molten material. Tides can be also in the interior of a planet, which has a fluid core under a hardened or solid crust. If you had a hollow ball, and filled it with thick syrup, when you spun the ball over you would

---

* This exhaustion of energy by tides may account for the slow rotation of the moon, making one moon day a month long.

create motion in the syrup. When the force in question is attraction extended over the body with the soft interior, tides are created. The force of tidal ebb and flow acts upon the planets in their motion, hastening or retarding their rotation.

In speaking of the building of our earth, its change from gaseous to a solid crust, and the transforming of this crust into continents, islands, seas, much of the surface becoming fertile, productive, or arable land, we noted that water had been a great earth-carver. Water had carved out shores, had channeled out river beds, had eaten out of straight coast lines the curves of bays and gulfs; water had carried sand, loam, alluvium; water had laid down miles of chalk or sandstone rocks. The tides on the earth during all these formative ages were busy workers; probably they were much higher, more masterful tides than now.

This work of water in cutting away or building up land still goes on. Those who lose land, house sites, by the fury of tidal wash, are likely to speak of tides as great robbers; but on the contrary, they usually give as much as they take, building up with one hand what they tear down with the other. At Cape May Point at very low ebb one often sees over a quarter of a mile beyond present water-mark the traces of wells and houses which were once upon the shore, with a considerable strip of land between them and the level of high tide. Now the land which they occupied is all under water and sea sand, while the coast line is nearly half a mile farther inland. At Atlantic City the same thing has occurred, though there is land there which has been captured and returned by the sea several times.

Some one has made the calculation that the gnawing and wash of tides will completely eat up the British Isles in the course of five billions of years. What will happen at a date so remote is probably of very little consequence to any one now living in England.

On the other hand the rivers seem to be combating with the tides, the tide's overflow, rendering salt and to a large degree worthless, much low-lying land at the mouth of great rivers. The rivers bring down from the uplands of the interior a very great quantity of rich soil, which is deposited on the low-lying land near the banks. As much as two or three inches of this good earth will be deposited in the course of a tide. In about three years seven feet of arable soil has been laid down in some localities, and the rivers Trent and Ouse, which are famous as carriers of solid matter, have built up near the coast thirty thousand acres of exceedingly fertile land. In our own country the Mississippi carries down from the far Northwest, the headwaters of the Missouri and other rivers, more than three millions of tons of good soil each year, and where this is spread about on the line of the Gulf good land appears where long ago was only the salt marsh.

Against this laborious carrying and building of the tireless rivers we may set the fury of sudden great and terrible tides when sun and moon and storms unite their force, swelling over islands and sweeping away in a few hours all vegetable and animal life, leaving them wastes of bare, tide-borne sand, riven, gullied, desolated by the immense fury of the waters.

The rivers, in bringing down the enormous amount of earth, leave the interior highlands denuded, and as this process goes on illimitably we may note that the work of the rivers is transference merely. What it places in one locality it tears away from some higher locality, placing the material where the tide may eventually eat it up.

CHAPTER XVIII

"Many a night I saw the Pleiades, rising through the mel-
low shade,
Glitter like a stream of fire-flies, tangled in a silver braid."

How often not only as children, but when of a
mature growth, we have wished for one of those
wonderful genii told of in the "Arabian Nights,"
bound to do the bidding of him who released the
genius from under the seal of Solomon. What
errands there are beyond human accomplishment
which such spirit could perform for us! How frequently, on starry nights, when we are ravished by
the splendors of "the vault studded with inextin-
guishable fire," we have wished for a messenger to
send to the far-off stars. Now we are fortunate, for
not only has one afrite been on such an errand, but
three have reported their discoveries.

What are the names of these wonderful servants
that have voyaged into space for us? Afrites Telescope, Chemistry, Mathematics. Afrite Mathematics
discovered Neptune and the asteroids. He reports
that the stars are more distant than we can imagine.

176

The numbers which he uses to express their far-offness are too great for our comprehension. These stars are called fixed stars, because from age to age their general position remains the same. Doubtless they are all moving swiftly, carrying their systems with them, but so great is their distance that only careful calculations can detect their change of place. We know that ships far out at sea seem motionless as we watch them from the shore, although they may be moving a number of miles an hour.

When we look at these stars they sparkle or twinkle. The planets shine upon us with a serene, steady light. The fixed stars are light-givers as the sun is, not light-receivers, as the planets.

The size of the stars varies, and for convenience in study they have been divided, according to size, into twelve classes. Six classes of the largest are visible to the naked eye; the others can be seen only by telescope. The reason of this variation in size is not always a real difference in the magnitude of star and star; it is often the distance that causes apparent difference.

No doubt there is a real difference in magnitudes, as is said in the Bible: "There is one glory of the sun; and another glory of the moon; and another glory of the stars: for one star differeth from another star in glory."

12

Most of the fixed stars are larger than our sun. Sirius is supposed to be as large as eight suns like ours. Vega is as large as thirty-eight suns.

Mathematics has been able to weigh the stars, but not to measure them accurately. This magician hints that there are nine thousand millions of star-suns in space. He reminds us that our sun is a star, and that all the light we have is starlight. We talk of "moonlight," "sunlight," and "starlight"—it is all starlight. Our sun is a star, our moon shines by reflection of that star.

As to distance, one tries in vain to realize it. The nearest fixed star is trillions of miles off. Light travels at the rate of one hundred and eighty-five thousand miles a second, yet so far off are the stars that it takes their light from three and a half to many thousand years to reach us. If to-day one such star suddenly perished, for a thousand years the light that has already left it would be streaming to us.

What does the Telescope report to us? He is younger than Mathematics; Mathematics is the afrite of certainties of fixed laws. The Telescope varies his reports with his growth and his adjustment. He has made many excursions into space. He says that indescribable beauties reward his search. He has found out that stars which seem to lie near together seem so from their great distance from us, and really are far apart. He has had amazing revelations of size and brightness.

Herschel was one day looking through his telescope, when Sirius, a star of the first magnitude, was brought into the field of vision. The splendor was as the dawn of day ; dazzled by the glory the astronomer was obliged to turn away his eyes, as if he had tried to look upon the sun. Sirius is the most brilliant of the stars, and is a million times farther from us than the sun is. Sirius moves at the rate of a thousand miles a minute, and shows variations or oscillations that suggested that some large body near him must be attracting him strongly.

After search, this body was discovered by a young man named Clark, who, with his father, was trying a great glass for the Chicago University. Mr. Clark turned his new glass toward Sirius, and cried out, " Father, this star has a companion !" Thus the disturber of Sirius was found. This companion of Sirius is seven times heavier than our sun.

The telescope has learned that some stars are variable. Algol, a star of the second magnitude, shrinks to a star of the fourth ; then waxes again to the second, and so in ceaseless alterations.

Herschel discovered that there are stars which to the unaided eyes appear as single points of light, but through the telescope are seen to be double or multiple. A list of eighty double stars has been published. There are stars which seem to be double which may really not be connected, yet there are true double or multiple stars, and these vary in colors. Mathematics came to our aid to explain these double stars. Castor is a double star; the Pole star has a little companion, while our red star, Beltegueze, is one of a triple set.

The telescope afrite found red, blue, white, yellow, and green stars. He also reports what are called temporary stars—stars that disappear, stars that appear where none were seen before. Another discovery by these two afrites is, that no stars are fixed in the sense of being moveless. "Systems revolve about systems; suns about suns." The whole splendid retinue seems sweeping toward some common centre, in the direction of the constellation of Hercules, which, thus far, seems to be the centre of stellar motion.

What has Chemistry done for us?

Mathematics set off with scales in one hand, and a measuring rod in the other. Chemistry took with him a prism. His object was to find out of what worlds were made. He had a list of gases—he weighed, analyzed and compared. He came back and hinted that the colors of stars might be accounted for by the stage of development as worlds which they had reached; it is suggested that white stars come next after nebulæ, and yellow stars are a little farther advanced in density; with red still farther on. When Chemistry made that suggestion science looked doubtful, and said : " Wait a while and investigate before you announce that."

"At least," replied Chemiso, " I have found out for certain that all these spheres are made of about the same material. I have also been assured that the sun is not likely to burn itself out; that there are various ways of renewing its energy, restoring its heat and light."

Perhaps we should add to the three afrites already mentioned, a fourth, a modern, who may be capable of greater and greater achievements. This is the Camera. It is only very recently that the camera has been set at work in the cause of astronomy, and already it has found scores of asteroids, and has shown us discs of some of the far-off stars, discs which even through the telescope could not be grasped by the human retina.

Of all the six thousand stars which are visible to the naked eye, only about one in a hundred show any variations, but the camera has shown to us clearly the periodic changes of certain beautiful star clusters where six or ten in each hundred are variable. One of these star clusters where numbers of orbs are variable is in the constellation of the Hunting Dogs, and another in Sirius. These stars having been repeatedly photographed, when the negatives were compared it was found that the stars varied in the photographs. As might be expected, the observations and photographic work which detected these changes were possible in the observatories in the clearest atmosphere, and Arequipa, Peru, is the station where the best work of this kind has been done.

These great star clusters are so packed, as we may say, with stars that at their centres the star images crowd and overlap, this not because they really crowd upon each other in their celestial stations, but from number and immensity of distance they appear to do so, until these groups of stars revolving at vast distances each from each, appear on the photographic plate or through the finest telescopes to be hazy and crowding masses of fire. At such distances no stars appear to be of great magnitude; all are small. Now it is found that these stars change, expanding or contracting their light, so that at one time they may be half or double as large again as at other times, or may seem to shrink from their usual size. It was formerly supposed that only a very few such variable stars existed; now they are discovered to be numer-

THE AFRITES OF SCIENCE

ous. The reason for their changes has not been fully ascertained; several suggestions have been made. As all these stars are suns, and are no doubt much greater than our sun, it is possible that they change in size and light-giving by outbursts of chemical activity. Our sun has been known to shoot out spires or jets of burning material, which are twenty or forty thousands of miles in height, and vary several thousands of miles in a few hours. Possibly these variable stars have periods of excessive combustion, when prominences of such immense extent and intense brightness leap up upon their entire surface that they increase their magnitude to one-half as much, or even double it, and then by the transient subsidences of so great activity the apparent dimensions are again reduced. To such outbreaks as these in the far-off variable stars the excitements or corruscations of the surface of our sun are mere miniature. Professor Young suggests that an explanation of some variations of stars can be found in eclipses which are taking place when the particular star is under observation; or the intense access of light may be due to collisions between the atmospheres of stars brushing past each other in their flight through space; and this theory is given a certain probability by the fact that there are the most exhibitions of variation in stars that are near the centre of clusters where stars are very much crowded. No doubt the list of variable stars will be largely increased. Already the period of variation or disturbance can be predicted in some of these stars, as eclipses of the sun or moon can be predicted, but according to astronomers the majority of the variables are very antic and erratic in their behavior, and no one knows when they will " break out."

Small as is our knowledge of the fixed stars, it is yet immensely greater than it was less than a century ago, and now makes rapid progress, keeping pace with the wonderful improvements in the power and precision of telescopes, and with the application of new instruments, such as the spectroscope and the camera, to the study of astronomy. The achievements of the modern astronomy would have seemed a fantastic dream to the star-students of Galileo's day.

Once a telescope for astronomical use was a small instrument that could be easily lifted by the hands, and having a lens a few inches in diameter; now at Mount Hamilton, in California, there is a telescope with a length of sixty feet and a lens of thirty-eight inches clear diameter; the telescope of the Lick Observatory has a lens forty inches in diameter, the length of the instrument being sixty feet. Such a telescope is mounted on a ponderous pillar, and a mechanism so delicately poised and adjusted that the very slightest movements can be quickly and easily made. In some cases the needed movements are given by clockwork.

The telescope of the Earl of Rosse at Parsonstown is mounted between two pillars of masonry, and the tube of the instrument is so "large that a tall man can walk through it without stooping." Yet we need not suppose that all the work, or the greatest work of astronomy, has been done by means of these enormous instruments—the moderate sized telescopes, which are brought to a great optical perfection, have often been used in the most fortunate observations. By all these methods of modern science, by these instruments of the nineteenth century, we make our harvests of knowledge among the far-off stars.

CHAPTER XIX

"A glittering star is falling
From its shining home in the air."

WE would do well to be wide awake to-night, for there is to be a "meteoric shower;" probably it will begin by nine o'clock, and what are popularly called "shooting stars" will be plenty. A bright light flashes along the sky and fades, then another, and another. Some one cries out in the darkness of the chilly November night, "I see a falling star; one, two, three!"—the rain of harmless fire has commenced.

These are called stars, but are not stars at all; they are meteors. If they are not stars, why do they shine? They shine because they are hot, red hot, white hot. Why do they go out? Let us discuss affairs around and about this question before we answer it directly. Does not a bullet or a cannon ball become heated when it is fired from a gun? We have picked up bullets that are very warm, just after they have been fired at a mark: they have been thus much heated, simply by passing swiftly through the air for the short space that separated the marks-

186

man from the mark. The rubbing or attrition against the air heats them. Meteors move so much faster and farther than shot or balls that they become heated incomparably hotter. In the emptiness of space they whirl about the sun, each safe in its own track, until that track sweeps it into the earth's attraction. When, several hundred miles from our earth, it enters our atmosphere, friction heats it and presently its visibility begins, its destruction may be near. Finally many of them are by their intense heat expanded into fluid, then into gas, and burst into atoms, as an over-blown bubble bursts. When they burst we lose sight of them; we say "they go out" or "have fallen to the earth." Millions of them are held as fine dust, ever drifting about in the upper layers of our atmosphere. Some of them fall upon the earth before they lose their condition of stones. For a long time these meteors were regarded as a fashion of accidents—unexpected, fortuitous happenings; but nature has no accidents, all moves by steadfast laws. It was finally noted that at certain periods of the year these meteors were especially numerous, so that they were said to fall in meteoric showers. The time of these showers being known, it was less difficult to reason of the why of their occurrence.

While the two great shower periods give us an exhibition of the greatest number of meteors there is scarcely a night in the year when one or more of these bright bodies does not flash in a trail of fire across the heavens. These are laggards from the one grand army of meteors or scouts sent hurrying in advance of the other grand army. The investigation of these meteors is of much interest. We have talked about enormous heavenly bodies. Even the smallest of the asteroids and satellites which we have spoken of is large in comparison with these meteors. Meteors vary in dimensions from many tons weight, or many pounds weight, to the size of eggs or little pebbles, and some astronomers suggest that there are plenty of individual meteors no larger than grains of sand. " There are more meteors in space," says one, " than there are fishes in the sea."

Specimens of meteoric stones are to be found in most museums. Several weighing over fifty pounds are known. In the fall of meteors so large as these, trees are known to have been cut down or broken, and the stones have been found buried from one to three feet in the soil by the force of the fall.

These meteoric stones are not cold shooting stars that have expired upon the earth. The meteoric stones have probably a different origin, though exactly what that is, is yet in doubt. Late as we live in the world's history there is yet plenty to be learned.

We are told that there are inconceivably many meteors, and that they move in great " shoals," as they are called. They obey the laws of the solar system; the shoals are widely scattered, yet the members are held in company by attraction, traveling in an ellipse about the sun.

The strongest telescopes cannot discern the individual bodies, until drawn by the attraction of our earth they leave their wonted orbit and rush toward her surface. Thus entering our atmosphere we call them " shooting stars," from their brilliant light and swift descent.

The velocity with which these meteors approach us is great, often more than twenty miles a second. Near the surface of the earth our atmosphere is more dense, and that high rate of speed cannot be maintained. Far up in space there is no atmosphere to hinder their onrush in the direction of any planet which has laid hold of them by attraction.

The entrance of a meteor into our atmosphere is like firing a bullet into water. The speed is checked by the density of the medium which it enters. Yet the speed is still so great that the friction of the atmosphere makes the meteor red hot, white hot, and finally many of them are so hot that they are converted into vapor, and vanish away in invisible gases.

Are we inclined to ask how all that heat can be developed by such a small affair as friction? Friction is a very great affair. When just the little friction we can arouse by rubbing a knife-blade on a piece of cold iron, or rough carpet, will make the blade so hot that it will burn one; when we can rub two sticks together until they take fire, we might guess what heat would be excited by matter moving twenty miles in a second.

Some shooting stars have left famous records of their size and brightness. Some have been seen for several seconds by many people scattered over a large extent of territory. One seen in England in 1869 had a long train of light behind it that was seen for fifty minutes. This appeared like the tail of a comet and was, no doubt, a portion of the meteor reduced to burning gas by the heat of friction.

Various wonderful star showers have been noted and recorded in history. These records finally directed attention to the time of year in which the showers occurred, and this led to further investigations. During the last one hundred years various facts in connection with shooting stars have been arrived at.

There are within the bounds of our solar system

certain vast shoals of these small meteoric bodies.
They are inconceivably numerous and lying from
one to several miles each from each. A shoal may
stretch over hundreds of thousands of miles. Each
shoal of these minute bodies pursues an elliptical
path about the sun, just as do the separate
planets.

On they sweep in their broad track, and if nothing interfered with their journey they would pursue

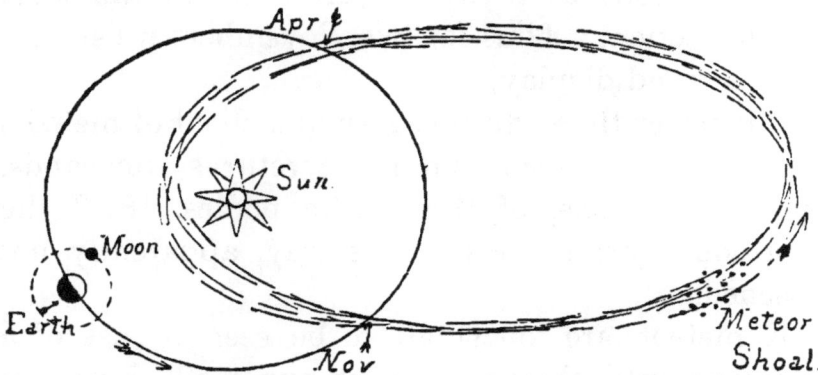

it again and again for ages. But something does
interfere. The crowded shoal of little bodies crosses
the orbit of our earth, and drawn by her attraction
myriads of meteors desert their ranks and rush to
her. In November and April, as may be seen by this
sketch, our earth crosses a meteoric shoal orbit, and
thus, knowing when the orbits cross, we can predict
the star shower. There are also, as has been said,
laggard meteors, drifting behind their train, which

may be caught by our greedy earth on other occasions.

Several such shoals of meteors are known and named. One is called the Leonids. This one is so large that it does not make its circuit in one of our earth's annual evolutions. It only intersects the orbit of the earth once in thirty-three years. So enormous is the extent of that meteoric host that our earth rolls by it twice before it passes the point of intersection. Thus for two Novembers we should have a grand display.

Whenever the earth meets such a shoal of meteors she draws to herself, and never returns, thousands, may be millions, of these little bodies. Still the enormous stream rolls on its way, apparently not lessened.

If meteors are too small to be seen by use of a telescope until they come near our earth, how can their path be found out?

Although the meteors are invisible, great comets are very visible. It has been found that the known track of a great comet is identical with that portion of the track of a meteoric shoal of which we are assured, as the Leonids and some other shoals. This relationship between comet tracks and meteor paths has been observed more than once. This leads not only to the discovery of the orbit of the shoal and the return to intersect the earth's path, but suggests that there is some close and singular connection between star showers and comets.

Thus far we have spoken of shooting stars or great meteors. We are told that some of these reach the earth as meteoric stones. The meteoric stones which are mentioned as having fallen to the earth are not the same class of bodies as shooting stars. Meteoric stones have been known since stones began to be studied, but it was long denied that they fell from the clouds. "It is impossible. There are no stones in the clouds," folks said.

Within this century the celestial origin of meteorites has been admitted. They do fall upon our earth. Some of them are almost pure iron, and these have been found on the earth's surface, far from any iron deposits. Many meteorites have been seen to fall. The fall is accompanied by a loud, rushing noise, and an explosion. When picked up soon after their fall they are warm or hot. In shape they are rough and irregular, like broken fragments from larger bodies. Gold, carbon, helium, and other materials are found with the iron in these stones.

There are many theories about the origin of meteorites. Some fancy that they are cast out by volcanoes upon some other planet; others think that they come from volcanoes upon the earth, in the

13

earlier ages of earth-building, and have been whirling around in our atmosphere ever since, but have finally tumbled back to earth as their motion slowed up. All these are but unproven theories. A meteoric stone appears to be a very simple affair, but it is a something which we know very little about.

The infinitely great and the infinitely little are alike beyond the grasp of human comprehension. The problem of the outposts of this wonderful universe eludes us; we can study only those nearer parts within telescopic range. So with these very small parts of our system, the tiny meteors, small as dust, we cannot pick them up with eyes or instruments; but as with that part of greatness which lies near us, and we can see and reason about, so with that part of smallness, these little vagrant bodies which become near and visible, we may study them and argue from less up to more. As of all other bodies of our system except the sun, we say of the meteors they are not self-luminous, but neither is their brightness reflected as that of the planets; it is brilliancy gained by rapid motion, the converting of that motion into heat. Our earth moves about thirty-two thousand yards a second, and the meteors move about thirty-nine thousand yards in a second; when a shower of these meteors meets the earth's atmosphere the combined speed is very great, and the shock produces a tremendous heat ; the meteor may be melted or volatilized at once, or it may be merely heated very hot, and go on through our atmosphere, increasing its heat. Very likely the greater part of the meteors which come within our atmosphere— and they are computed to be forty-six billions yearly —reach the earth in the form of a fine deposit of dust, and are thus slowly adding to its mass.

Various interesting experiments have been made with meteoric stones. Mr. Lockyer found that by heating them to luminosity he could obtain by the spectroscope evidences of matter such as we find in the spectra of nearly all heavenly bodies, and from this he concluded that all stars, nebulæ, comets, are closely related to meteors, and grasping one of them we hold part of the matter of which all the splendid worlds of space are made.

A great part of what is said of meteors is entirely, and some of it idle, conjecture. For instance, it is said that they intercept the sun's heat irregularly, and so change our earth temperature, making fitful seasons ; that they cause zodiacal light, and also auroras, and that they affect the moon in the same way that they do the earth ; that they also are the fuel of the sun, falling into his blazing fires and re-supplying them with combustibles. The absurdity of some of these theories has been abundantly proven, and others are yet under discussion.

CHAPTER XX

"Comets, imparting change of times and states,
Brandish your crystal tresses in the sky!"

IT seems that shooting stars, meteors and comets
are pretty closely connected, whether we consider
their motions or material; indeed, one or two noted
astronomers have suggested that these "shoals of
meteors" are only comets that have gone to pieces;
and that at least some comets are "swarms of stones,"
revolving swiftly, kept closely together, whirling
around the sun, held by his attraction. Chemistry
shows us that comets and meteoric stones seem to be
of the same material in different states.

For three months, in 1858, Donati's comet was
visible in its vast distance—small, pale, hanging like
a bit of floating nebulæ against the sky, while science
and newspaper gossip chronicled its every change,
and it was watched with awe or terror. Comets have
regular orbits much more elliptical than those which
the planets follow. This long ellipse brings them
very near the sun in some places, and very far from
the sun in other parts of the orbit. The velocity

196

of the comet is enormous when near the sun, when far away it moves slowly. As the comet of 1858 in September neared the sun in its path, a vast and splendid tail unfolded; a tail supposed to be three millions of miles in length. The nights were dark, and this great flaming thing sped across the heavens, wrapped in a strange glory.

Another comet appeared in 1882. Early in the morning before the dawn one might watch on our Eastern coast, above the waste of the Atlantic, that long train of light, that glowing head, rushing sunwards. This comet was so intensely bright that it was even seen in the daytime. Three things about it made it famous : it was photographed ; the spectroscope analyzed it, and it was found that it had sodium, iron, and carbon among its materials; and it finally approached nearer the sun than any comet that had been previously observed.

Comets approach nearer the earth than stars are found. Some of them return at short intervals, and seem to belong entirely to our system. One or two are known to have part of their orbital path beyond Neptune. Other comets appear to us once, among our planets; they seem to come from far-off stellar space. They are visitors, coming apparently for the first time, and retreating without any distinct promise that mankind will ever see them again.

We know that comets are composed of very thin material, of the most delicate gases, for sometimes they pass between us and some of the stars; and then the stars can be seen about as clearly through the volume of a comet as if the comet were away. Even a thin nebulæ can be seen through a comet's tail. Arcturus showed as bright as ever through Donati's great comet. Any gas or air which we know anything about in this world would refract or bend the rays of a star seen through it, and so make it change its apparent place. Comet stuff has no such refracting power; while we speak of it as gaseous, we know that it is far less dense than any gases that have been dealt with in this world.

Comet stuff being of such small density we understand that their weight must be but small indeed in proportion to their bulk. Therefore their attractive power is very small. If this were not so they would be dangerous visitors in our solar system. In their onrush toward the planets they might derange the planets' motions, even drag them from their orbits. Instead of this the planets seem to be uninfluenced by these wild intruders.

There are more comets than planets. " Comets are as multitudinous as insects in the forests, or as fishes in the sea," says one astronomer. Their orbits are very eccentric; not only that, they zigzag in and out

about their orbital track, and dash about at all sorts
of angles as capriciously as little dogs at play.

Comets have been named after astronomers who
have calculated their orbits and predicted their return. Thus we have Donati's, Halley's, Encke's,
Biela's comets, and others.

Comets have strange and varied shapes. Some are
curved like swords, or bent into bows. We associate
tails with comets, but many are tailless, being mere
burning heads wildly floating about. Some, on the
other hand, have two or three tails. Cheseaux's
comet had six tails.

The tails of comets are always turned from the
sun, and the nearer the comet is to the sun the
greater the expansion of the tail. As the comet recedes from the sun the tail again shrinks.

The early history of Halley's comet has been
learned from records kept by the Chinese. This
comet is to blame for many of the silly superstitions
about the direful omens of comets, and that they
appear as presaging human disasters. It happened
that this comet appeared several times on the eve of
more or less great events, and was therefore accused
of foretelling these events. It is very easy to be
mistaken.

Halley's comet has been within visible distance of
the earth twenty-six recorded times. In 840 A. D.,

it frightened Louis LeDébonnaire, King of France, to such an extent that he set himself to fasting, praying, building churches and convents. Two years later he died, and every one was sure that the comet had been sent as a harbinger of his death. Surely a great messenger for a small errand.

Back came this same comet the very month that William the Conqueror took possession of England. In 1455 the Turks and the Crusaders were engaged in terrible battles, and the Turks threatened to overrun Europe. Again this same comet appeared. The Turks claimed that the comet had come to aid them. However the Christians won the day, and all the church bells were rung for joy of the victory.

When Halley's comet made later returns people had learned that the great sky vagrant had no connection with human events.

Newton and other astronomers have predicted that eventually the sun would draw into himself all the comets, thus renewing his store-house of light and heat. One remarkable fact about comets is that they have carbon in their composition, and carbon is a material usually associated with organic life. It is also a notable fuel, and many comets are doubtless to end by entering the globe of the sun, thus possibly returning to their source "as life to the bosom of Braham."

Encke's comet presents many peculiarities, and to the mathematical astronomer is, perhaps, the most interesting of all the comet tribe. For some reason the orbit of Encke's comet is constantly becoming smaller and rounder. It has always been among " the short period comets," and was the first of these noted. Its time has been three years and four months. Ordinary observers pay very little attention to this especial comet, for it is so faint and small that it can only be beheld through a telescope, and then is undefined and misty, like a smoke wreath drifting out and fading upon the air. Now it is seen that this pale comet is moving in a spiral, winding constantly inwards toward the sun.

It is estimated that in a thousand or fifteen hundred years Encke's comet will tumble into the great central fire about which it is now circulating with the infatuation and rashness of a moth about a candle. Before that happens some catastrophe may overtake and explode or dissipate the entire comet, as Biela's comet was supposed to have come to wreck; or possibly some new influence may turn it about and send it far away, again to reapproach the sun in an inward spiral path.

The best explanation of the curious changes in the course of Encke's comet is that it meets and is held back by the attraction of a swarm of meteors; the meeting of such a swarm, and the retardation of the comet, would cause the orbit to shrink, and also while the period was shortened the speed of the comet would accelerate. Encke's comet is supposed to be two hundred thousand miles nearer the sun than it was at the beginning of this century.

If comets were not subject to the universal laws of gravitation their paths could not be calculated, and their return accurately indicated. The path of a comet was demonstrated by Newton to be not an ordinary ellipse, but a parabola, which is an extreme form of ellipse; until his day the erratic orbits of comets had never been reduced to geometrical form. The moment a path of regular form was discovered, then the comet's whereabouts on that path at any given time could be calculated. A comet is known rather by its path than its appearance, for the form of a comet may change. As we have seen, its tail or tails expand and contract, and also comets lose their tails, the burning matter forming them being dissipated or reabsorbed into the head.

At any hour or at any season comets may appear in any part of the heavens, and wherever they are, wonder and admiration will follow their wayward paths.

These comets fill us with awe and an absorbing interest; the grandeur of the universe seems con-

stantly to grow upon the mind as it strives to follow
one of these flaming pilgrims along his mighty track ;
the soul is upborne into the region of higher and
higher things.

> " The stars are forth, the moon upon the tops
> Of the snow-shining mountains—beautiful !
> I linger yet with nature, for the night
> Hath been to me a more familiar face
> Than that of man ; and in her starry shade
> Of dim and solitary loveliness
> I learned the language of another world."

THE END

ETIQUETTE

BY AGNES H. MORTON

Author of "Letter Writing," "Quotations," etc.

Cloth Binding   50 Cents

SOME manuals of etiquette treat almost exclusively of "state occasions." Their instructions are chiefly such as only people of large wealth and abundant leisure have any occasion to follow, and are of little more value than a fairy story in guiding the daily conduct of the average social circle.

Another class of etiquette books is the commonplace compilation of sundry rules, often illiterate in style, and of doubtful authority. Such are misguiding to the ignorant, and promptly rejected by the intelligent.

Both of these classes and manuals are obviously inadequate to the needs of the great mass "who dwell within the broad zone of the average." For this large class, a book that gives information as to the essential points of correct behavior in social life, —points equally applicable to the rich and to the poor,—is the ideal manual. Such a book is the present volume. While it gives correct usage as demonstrated in the highest circles in the land, it also does more than this; it adapts these same principles to the life of the most unpretentious circles, constantly illustrating the maxim that the quality of courtesy is invariable.

Sold by all booksellers or sent, prepaid, upon receipt of price.

THE PENN PUBLISHING COMPANY

923 Arch Street

LETTER WRITING

BY AGNES H. MORTON

Author of "Etiquette," "Quotations," etc.

Cloth Binding   50 Cents

THIS admirable manual contains Suggestions, Precepts, and Examples for the Construction of Letters, and altogether is the most intelligent and thoroughly literary work on the subject ever offered to the public. It is from the pen of a skilled writer, who for several years filled the chair of Literature and Criticism in one of the leading educational institutions of the country.

The book exactly fulfills the promise of its admirably chosen title. Its suggestions are pointed and practically helpful; its Precepts are correct, and are clearly and attractively stated; its gracefully composed Examples are true to the character of the correspondence which they severally illustrate, and are accompanied with terse explanatory remarks.

Its object is to assist inexperienced persons to develop their talent for correct and graceful-letter writing. This gratifying result it will accomplish not so much by adhering to the numerous forms of letters applicable to all conceivable business and social occasions, as by following the excellent suggestions and directions with which the work abounds for the writing of original letters.

Sold by all booksellers or sent, prepaid, upon receipt of price.

THE PENN PUBLISHING COMPANY

923 Arch Street, Philadelphia

www.ingramcontent.com/pod-product-compliance
Lightning Source LLC
Chambersburg PA
CBHW021942220326
41599CB00013BA/1491